SpringerBriefs in Earth System Sciences

Series editors

Gerrit Lohmann, Bremen, Germany
Lawrence A. Mysak, Montreal, Canada
Justus Notholt, Bremen, Germany
Jorge Rabassa, Ushuaia, Argentina
Vikram Unnithan, Bremen, Germany

More information about this series at http://www.springer.com/series/10032

Manuel Enrique Pardo Echarte
Odalys Reyes Paredes · Valia Suárez Leyva

Offshore Exploration of Oil and Gas in Cuba using Digital Elevation Models (DEMs)

 Springer

Manuel Enrique Pardo Echarte
Exploration Scientific-Research Unit
Centro de Investigaciones del Petróleo
   (Ceinpet)
El Cerro, Habana
Cuba

Odalys Reyes Paredes
Department of Geophysical Interpretation
   and Structural Geology
Centro de Investigaciones del Petróleo
   (Digicupet)
Habana
Cuba

Valia Suárez Leyva
Department of Geophysical Interpretation
   and Structural Geology
Centro de Investigaciones del Petróleo
   (Digicupet)
Habana
Cuba

ISSN 2191-589X   ISSN 2191-5903   (electronic)
SpringerBriefs in Earth System Sciences
ISBN 978-3-319-77154-0   ISBN 978-3-319-77155-7   (eBook)
https://doi.org/10.1007/978-3-319-77155-7

Library of Congress Control Number: 2018934417

Printed on acid-free paper

This Springer imprint is published by the registered company Springer International Publishing AG
part of Springer Nature
The registered company address is: Gewerbestrasse 11, 6330 Cham, Switzerland

Foreword

The largest volumes of geological and geophysical information in oil and gas exploration are obtained from seismic, which is, in turn, the most expensive geophysical method. The authors aim to show how the more economical Digital Elevation Models (DEMs) provide a rapid form of preliminary offshore oil and gas exploration and are useful for focusing areas to be raised with other complementary unconventional methods, in order to propose, finally, the more expensive volumes of detail 2D–3D seismic.

The application of DEMs and other unconventional geophysical methods for the detection of light hydrocarbons (HCs) microseepage on the gasopetroliferous occurrences is based on the formation of columns of mineralized rocks above the deposits, called "chimneys." Its origin is due to the vertical migration of light HCs and to the oxidation of these by microbial action, which leads to the change in pH/Eh of the system by creating a reducing environment. This change results in the corresponding to the mineral stability, with the consequent secondary mineralization of calcium carbonate and silicification; denser, erosion resistant and resistive surface materials. This results in the formation of positive residual geomorphic anomalies and resistivity maximums. In addition, changes in pH/Eh precipitate other magnetic, iron and sulfurous minerals, which increase magnetic susceptibility and polarizability in the most superficial sequences, with the appearance of consequent magnetic and Induced Polarization (IP) anomalies.

In different scenarios, it is shown how the use of the DEM (90 × 90 m) provides a quick way of preliminary offshore oil and gas exploration:

- The detection of subtle positive geomorphic anomalies, possibly related to light hydrocarbons microseepage on oil and gas targets, is exemplified in Block L of the Central East-North Cuba sector of the Exclusive Economic Zone (EEZ) of Cuba–Bahamas–Haiti. The data of the residual DEM at 500 m are integrated with the geological, seismic and IP information, which represents, for the first time, the use of this type of anomalies for exploratory purposes in marine territories.

- The use of positive residual geomorphic anomalies of different amplitudes in the Northern Cuban Oil Strip (NCOS) of the Habana–Matanzas region is validated with the objective of establishing regularities in their occurrence and revealing new areas of petroleum interest in the regions where there are known producing fields. The data of the residual DEM at 500 m are integrated with the information of the exploitation wells and the mapped seismic structures.
- The first test of a prospective interpretation methodology is shown where the information on the location of possible producing structures, based on the positive residual geomorphic anomalies, with the corresponding evaluation of the possible fluid saturation, from the use of seismic attributes, is combined. This allows revealing new areas of oil interest. Positive residual geomorphic anomalies obtained from the residual DEM at 500 m are used as well as the seismic attributes "Relative Acoustic Impedance" and "Genetic Inversion," from which volumes of assisted properties are generated with data from wells, responding to the presence of reservoirs.
- Morphotectonic regionalization is carried out in the Central East-North Cuba sector of the Cuba–Bahamas–Haiti EEZ, which allows to identify four regions where the tectonic environments of Platform (Remedios–Bahamas), Foreland Basin, and Thrusted Belt are recognized. An evaluation of the degree of similarity between the section of the morphotectonic scheme belonging to the easternmost block and the morphotectonic section of the Seboruco–Varadero sector of the NCOS in Habana–Matanzas (one of the most productive sectors of the same) is done. The evaluation allows arguing, a priori, the gasopetroliferous perspective of the most eastern sector of Cuba.

A group of authors with a long history in geology and geophysics, distinguished as specialists in the search for oil and gas in Cuba, are participating in this research. Are they:

Valia Suárez Leyva, with 25 years of experience in scientific research, is a graduate of Geophysical Engineering at the Plekhanov Institute, St. Petersburg, in 1990, with a PM degree in Applied Geophysics at ITC, Enscheda, the Netherlands, 2001. She belongs to the Petroleum Research Center (Ceinpet), Digicupet. She is an Assistant Professor at the Habana University. It has a group of publications and participation in national and international events.

Geophysical Engineer Odalys Reyes Paredes, with 27 years of experience in scientific research, graduated in 1990 at the José Antonio Echeverría Polytechnic University, Cujae, La Habana, with an M.Sc. degree in Applied Geophysics, 2013, in the same institution. She is an Assistant Professor at the José Antonio Echeverría Polytechnic University, Cujae, and also is part of the collective work of Ceinpet, Digicupet. She has published numerous articles and participated in a series of national and international events with more than 20 works presented.

The main author, Manuel Pardo Echarte, with more than 40 years of experience in scientific research, obtained his graduation in Geophysical Engineering at José Antonio Echeverría Polytechnic University, Cujae, La Habana, in 1974, where he obtained a Ph.D. in Mineralogical Geologist in 1987. He is Associate Professor of this University and the University of Habana. It has a wide range of publications, including monographs: "Unconventional Methods for Oil & Gas Exploration in Cuba: The Redox Complex" (Springer, 2016) and "Oil and Gas Exploration in Cuba: Geological-Structural Cartography using Potential Fields and Airborne Gamma Spectrometry" (Springer, 2017). He is the author of the "Method of Redox Potential measurement in soils and its combined application with Kappametry for the purposes of geological prospecting" (2000) and of the Redox Complex (2004). He has presented numerous works in national and international events. He currently works in the Geology Unit of Ceinpet.

Habana, Cuba   Dr. Olga Castro Castiñeira
                                                          Head of Well-log Analysis Department
                                                          Centro de Investigaciones del Petróleo

Preface

Exploration–production operations in marine territories pose major challenges for the oil and gas industry sector. To deal with them, a fundamental change is needed in the way the marine oil and gas industry operates. The projects associated with these environments are best visualized as a single integrated effort of all possible disciplines ranging from exploration to production.

In this sense, it provides an overview of the major changes induced by hydrocarbons (HCs) affecting rocks and surface sediments and their implications for non-seismic exploration methods, particularly for marine territories. The bacterial oxidation of the vertically migrating hydrocarbons can directly or indirectly generate changes in the pH and Eh of the surroundings, changing also the stability fields of the different mineral species present in that environment. These changes result in the precipitation or solution and remobilization of various species and mineral elements, such that the rock column above an accumulation of oil becomes significant and measurably different from that of the laterally equivalent rocks, forming a type of "chimney." Diagenetic carbonates and carbonate cements are among the most common hydrocarbon-induced alterations. In the seas, carbonates can form like slabs and debris, large mounds and pillars, hard terrain or carbonate cements that fill the porosity, all of which give rise to a positive residual relief. It should be pointed out that the positive residual geomorphic anomalies could also have a lithological or structural nature not linked to hydrocarbon microseepage processes. The interpretative procedure chooses to obtain and evaluate the geomorphic anomalies related to the hardening of the surface rocks on the HCs deposits, determined from an assay test method with known gas–oil targets data, has been the Upward Analytical Continuation at 500 m.

At the Block L of the Central East-North Cuba sector (Cuba-Bahamas-Haití Exclusive Economic Zone), the use of a Digital Elevation Model (DEM) at 90 x 90 m resolution for the detection of subtle positive geomorphic anomalies (possibly related to hydrocarbon microseepage on oil and gas targets) is exemplified. DEM data are integrated with geological, seismic and Induced Polarization (IP) information. The results support the conclusion that DEM data provide a lower cost and fast offshore oil and gas preliminary exploration strategy. These data are useful for targeting areas to be raised with complementary unconventional methods such as Magnetic and IP surveys and, then, to propose the most expensive volumes to raise with the 2D–3D detail seismic. In the particular case of Block L, the most intense geomorphic anomaly does not coincide with a local dome of any Upper Jurassic horizons (presumed reservoirs), suggesting a possible link to a "Gas Chimney" related to a local converging area of several tectonic dislocations inside the North Cayo Fragoso structure. This hypothesis would also explain the observed intense IP anomaly. A "Gas Chimney" seems to be confirmed by 2D seismic information, for which the vertical zones of light hydro-carbon seepage are evident as well as their exit to the oceanic bed in the form of small podmarks.

The use of the DEM to a sampling of 90×90 m for the detection of positive residual geomorphic anomalies possibly related to hydrocarbon microseepage on known oil and gas fields or production areas is validated in the North Cuban Oil Strip (NCOS) of the Habana–Matanzas region, Cuba. The data of the residual DEM at 500 m are integrated with the information of the exploitation wells and the mapped seismic structures. The 60% of the known offshore fields or production areas of the NCOS have positive residual anomalies; of them, 55% have amplitudes between 10 and 25 m and 22% are of amplitude greater than 50 m. All fields in operation are characterized by seismic structures. The 79% of the mapped seismic structures intercept positive residual geomorphic anomalies, of which 53% have amplitudes between 10 and 25 m and 27% are greater than 50 m. Mapped seismic structures (and reservoirs) that do not correspond to positive residual anomalies are located within a strip of increased positive values observed from land up to 1–2 km offshore in the Mariel–Matanzas Bay sector, where, because of the character of the residual DEM, it is impossible to detect isolated anomalies within it.

With the objective of revealing new areas of petroleum interest, a study is carried out in regions of the Northern Cuban Thrusted and Folded Belt where there are known oil-producing fields (within the NCOS), helping to establish analogies in unknown areas. Positive residual geomorphic anomalies, obtained from residual at 500 m DEM data (possibly linked to hydrocarbon microseepage on offshore oil and gas targets), as well as of the seismic attributes "Relative Acoustic Impedance" and "Genetic Inversion," from which volumes of assisted properties were generated with data of wells (responding to the presence of reservoirs), were used. The analysis of the coincidence of the geomorphic anomalies with the already drilled structures allows proposing the existence of a relation between its amplitude and the depth of the structures, noting a decrease of this with the depth of the drilled objectives. The response of the seismic attributes in question, linked to the presence of geomorphic anomalies in non-drilled zones, allows defining the possible extension of reservoirs in operation and new areas of interest not yet drilled.

From the DEM 90×90 m in its residual variant at 500 m, the offshore mor-photectonic regionalization in the Central East-North Cuba sector is performed. Four regions were identified (A, B, C, and D), where the tectonic environments of Platform (Remedios–Bahamas), Foreland Basin, and Thrust Belt are generally recognized. This regionalization was compared to the seismic-geological scheme for the easternmost region of the area, serving as a general validation, even though with a greater degree of structural detail. In addition, it provides a terrestrial gravimetric regionalization scheme, which allows proposing a scheme of tectonic zonation of the territory from its integration with the morphometric data. The results allow establishing a comparison between the anomalies derived from the seismic interpretation and the positive residual geomorphic anomalies of different amplitude, revealed by the residual DEM at 500 m. For the southeast of region C, the presence of small geomorphic anomalies with amplitudes between −40 and −10 m, and a smaller group with positive values in the first tens of meters, is established. Correspondence with seismic anomalies is around 50%. In region D, the predominant amplitude is 10–25 m, with an approximately 63% of correspondence with the interpreted seismic anomalies. An evaluation of the degree of similarity between the section of the morphotectonic scheme belonging to the easternmost block and the section of the morphotectonic scheme corresponding to the Seboruco–Varadero, most productive sector of the NCOS, is carried out. This allowed arguing, by simple analogy in the geological and morphometric data, the gasopetroliferous perspective of the most eastern sector of Cuba.

Habana, Cuba   Manuel Enrique Pardo Echarte
                                                                     Odalys Reyes Paredes
                                                                       Valia Suárez Leyva

Acknowledgements

We thank our institution, Centro de Investigaciones del Petróleo (CUPET-Investigación), for allowing us to publish partial information concerning various research projects and, particularly, Figs. 2.1, 2.2, 5.1, and 5.4, of its own scientific production.

We want to thank Fidel Prieto Castro for the confirmation of the different horizons of the Upper Jurassic, generated from the seismic interpretation with the Petrel software of Schlumberger (Figs. 2.5, 2.7, in Chap. 2).

We also want to thank for the partial or total revision of the manuscript and, for the correct observations to the same, of the following researchers: Dr. Osvaldo Rodríguez Morán, Dr. Evelio Linares Cala, Dr. Olga Castro Castiñeira, Dr. Reynaldo Rojas Consuegra, and M.Sc. Julio Gómez Herrera.

Contents

Abbreviations

AGC	Amplitude Gain Control
CBH	Cuba–Bahamas–Haití
CO_2	Carbon dioxide
DEMs	Digital elevation models
EEZ	Exclusive Economic Zone
Eh	Redox potential
EIUS	Exploration Information Unique System
EODC	Extraction and oil drilling company
FB (CA)	Foreland basin
GA	Genetic algorithm
GAs	Geomorphic anomalies
GI	Genetic Inversion
H_2S	Hydrogen sulfide
HCs	Hydrocarbons
IP	Induced Polarization
Ma	Millions of years
NCOS	North Cuban Oil Strip
NCTFB	Northern Cuban Thrusted and Folded Belt
NN	Neural Network
°API	Measure of density, by the American Petroleum Institute, compared to water at equal temperatures
pH	Degree of acidity or basicity of an aqueous solution
PreSDM	Pre-Stack Depth Migration
PWAI (IP)	P wave acoustic impedance
PWV (VP)	P wave velocity
RAI	Relative Acoustic Impedance
RCB (BCR)	Remedios carbonated bank
SWAI	S wave acoustic impedance
SWV (VS)	S wave velocity
TFB (CC)	Thrusted and folded belt

TVDSS	True Vertical Depth Sub-Sea
UAC	Upward analytic continuation
USTB	Unit of science and technology of base

List of Figures

List of Tables

Chapter 1
Theoretical Framework

Abstract Hydrocarbon-induced changes are measurable and mappable and have formed the basis of many different methods of surface exploration over the years. Unfortunately, understanding the complex physical, chemical, and biological processes responsible for these phenomena remains incomplete, with the result that these methods are viewed with skepticism and remain underutilized. The aim of this chapter is to provide an overview of the major changes induced by hydrocarbons (HCs) affecting rocks and surface sediments and their implications for non-seismic exploration methods, particularly for marine territories. It is proposed, moreover, to know about the interpretive procedure used to obtain and evaluate the geomorphic anomalies related to the hardening of the surface rocks on the HCs deposits. The bacterial oxidation of the vertically migrating hydrocarbons can directly or indirectly generate changes in the pH and Eh of the surroundings, changing also the stability fields of the different mineral species present in that environment. These changes result in the precipitation or solution and remobilization of various species and mineral elements, such that the rock column above an accumulation of oil becomes significant and measurably different from that of the laterally equivalent rocks, forming a type of "chimney." Diagenetic carbonates and carbonate cements are among the most common hydrocarbon-induced alterations. In the seas, carbonates can form like slabs and debris, large mounds and pillars, hard terrain or carbonate cements that fill the porosity, all of which give rise to a positive residual relief. It should be pointed out that the positive residual geomorphic anomalies could also have a lithological or structural nature not linked to hydrocarbon microseepage processes. The interpretative procedure choose to obtain and evaluate the geomorphic anomalies related to the hardening of the surface rocks on the HCs deposits (determined from an assay-test method with known gas-oil targets data) has been the UAC at 500 m.

Keywords Hydrocarbon microseepage · Hydrocarbon-induced alterations
Chimneys · Carbonate cements and diagenetic carbonates · Positive residual
relief · Upward analytical continuation

1.1 Introduction

The association of mineralogical changes and hydrocarbons microseepage has been recognized since the early days of oil exploration. Many of the early explorers noted the correlation of productive areas with paraffin, saline, or sulfurous waters, sub-surface mineralization of magnetite and sulfides, and topographic maxima. Such characteristics were fundamental, for example, in the discovery of about 70% of the oil fields of the coast of the Gulf of Mexico.

Hydrocarbon-induced changes are measurable and mappable and have formed the basis of many different methods of surface exploration over the years. Unfortunately, understanding the complex physical, chemical, and biological processes responsible for these phenomena remains incomplete, with the result that these methods are viewed with skepticism and remain underutilized.

The aim of this chapter is to provide an overview of the major changes induced by HCs affecting rocks and surface sediments and their implications for non-seismic exploration methods, particularly for marine territories. It is proposed, moreover, to know about the interpretative procedure used to obtain and evaluate the geomorphic anomalies related to the hardening of the surface rocks on the HCs deposits.

1.2 Synthesis of Changes Induced by Hydrocarbons in the Surface Environment

One of the more complete works on the changes induced by the HCs in the super-ficial environment is the publication of Schumacher (1996). In essence, it is argued that the bacterial oxidation of the vertically migrating hydrocarbons can directly or indirectly generate changes in the pH and Eh of the surroundings, also chang-ing the stability fields of the different mineral species present in that environment. These changes result in the precipitation or solution and remobilization of various species and mineral elements, such that the rock column above an accumulation of oil becomes significant and measurably different from that of the laterally equivalent rocks, forming a type of "chimney."

According to Price (1985), "chimneys" are rock columns above deposits of HCs that have different physical characteristics than stratigraphically equivalent rocks outside the HCs deposit. "Chimneys" are the result of the interaction of various bacteria with vertically migrating HCs (microseepage). Significant amounts of CO_2 and, to a lesser extent, organic acids, are produced by this interaction. These reactive chemical species significantly change the pH and Eh of the environments in which the bacteria are active and, change the mineral stability fields in the rock columns on the HCs deposits. These changes result in the solution and mobilization of some mineral species and the precipitation of other species. Several of these mineralog-ical changes have been observed in the rock columns above the HCs deposits: (A) Minimal radioactivity has been documented in clays and soils above HCs deposits, as compared to background values. (B) The sediment hardening documented above the HCs deposits is believed to be due to the precipitation of carbonates (from the microbially produced CO_2) and, to the silica cementation due to the decomposition of the clay into silica and alumina caused by high concentrations of CO_2. (C) Changes in iron/manganese-containing mineral species have been documented from alterations in pH/Eh and are believed to be due to bacterial oxidation of HCs in vertical migration. The consequent mineralogical modifications of iron/manganese result in anomalous magnetic and/or electrical potential compared to unaltered box rocks.

It was in the 1930s, in the USA (Rosaire 1938), when the "chimneys" were first described and documented, although at that time the term was not really used. It was observed that the sediments of some oil fields had abnormally high speeds because they were more hardened compared to those outside the structure. At that time, this hardening was attributed to the higher degree of cementation or mineralization in the sediments on the HCs deposits.

Also in the 1950s, in the USSR (Kartsev et al. 1959), soil carbonate–silicification based on enrichment of calcite, silica, and alumina was proposed on vertically escaping HCs deposits. The high concentrations of CO_2 due to the microbial oxidation of the HCs in migration caused in the soils not only the significant precipitation of calcite but also the cementation by silica, due to the decomposition of the clay in silica and alumina caused by high concentrations of CO_2.

Diagenetic carbonates and carbonate cements are among the most common hydrocarbon-induced alterations. In the seas, carbonates can form like slabs and debris, large mounds and pillars, hard terrain or carbonate cements that fill the porosity, all of which give rise to a positive residual relief. It should be pointed out that the positive residual geomorphic anomalies could also have a lithological or structural nature not linked to hydrocarbon microseepage processes.

The formation of secondary pyrite and other sulfides has been documented for many oil fields. Pyrite is the dominant mineral sulfide in the zones of alteration induced by hydrocarbons, but pyrrhotite, marcasite, galena, sphalerite, and native sulfur can be locally abundant.

In the 1970s and 1980s, in the USA (Donovan et al. 1974, 1975, 1979, 1981), cementation of near-surface rocks (mainly sandstones) by carbonates and the development of magnetite on oil fields were studied. The precipitation of such carbonated cements as a consequence of the oxidation of the HCs undergoing migration (microseepage), which leads to the production of CO_2, is then explained: It reacts with water to form the carbonate ion, which in turn causes the precipitation of various carbonate minerals. The precipitation of the calcite in the soils was not specific for the deposits of HCs; however, the ratio of the anomalous carbon isotope is observed in these cements, and they were specific for HCs deposits.

The presence of magnetic anomalies on oil and gas deposits has been observed for several decades. The same hydrocarbon-induced reducing environment that promotes the formation of uranium and pyrite leads to the precipitation of a variety of magnetic iron oxides and sulfides, including magnetite, pyrrhotite, and greigite. At that time (1978), the anomalous iron and manganese distributions in surface sediments on HCs deposits were explained from the passage of hydrocarbons and associated compounds such as hydrogen sulfide through the rocks cause a reducing environment. This environment reduces iron and manganese to lower valence states, with the consequent mobilization and elimination of the elements, or the precipitation of magnetite or maghemite, or coprecipitation of iron and/or manganese with calcite in carbonated cements on the deposits of HCs. The formation of magnetite, or other iron–manganese anomalies, is the basis of a number of different magnetic–electric geophysical exploration techniques including, but not limited to, aeromagnetic detection and Induced Polarization.

1.3   Synthesis of the Methodology to Evaluate the Geomorphic Anomalies

The morphometric methods used to evaluate the hardening of surface rocks on HCs deposits should consider the detection of very subtle or low amplitude geomorphic anomalies. For this, the regional–residual separation of the relief or Digital Elevation Model (DEM) corresponding to the seafloor is required. The separation of these two components consists of evaluating the regional characteristics and finding the local (residual) anomalies, subtracting the regional effects to the relief or corresponding DEM. The objective of the regional–residual separation is to obtain a map suitable to define the position, geometry, and amplitude of the local anomalies sought.

The methods of the regional–residual separation are many, but there are two generic categories: (1) graphical methods and (2) mathematical methods. Both have their advantages and disadvantages. Graphical methods are applied to profiles and/or contours maps. They have the problem that they are empirical and subjective methods, but also they are flexible in the estimation of those factors that can be taken into account, a priori, in the layout of the regional. By contrast, mathematical methods (sliding averages, polynomial surface fitting, derivation, digital filtering and spectral analysis, continuation or analytical extension up or down, etc.) minimize the subjectivity or predisposition of criteria, although there is always a certain degree of subjectivity, for example, in the choice of the mathematical technique to be applied. Hence, the separation of the regional and residual components is not a single solution, and it is often ambiguous and difficult to achieve.

In our case, the mathematical technique to be applied must satisfy, in its ideal character, the conditions to reflect, as objectively as possible, the geometric and amplitude characteristics of the local geomorphic anomalies (due to hardening of the surface rocks), as well as their real position in the plane. Within the most used mathematical methods for the estimation of the regional component is the polynomial adjustment. The regional is a function of smooth variation, analytically representable by the adjustment to the values of a generally small degree surface. This adjustment can be made by calculation methods such as least squares, but there will always be uncertainty as to the degree of the function to be fitted to the field in question. Zeng (1989) proposed a technique based on the analysis of continuations at different heights (analytic prolongation or continuation); in this case, it was for the gravitational field of Bouguer, to reduce this uncertainty.

After having tried in ground objectives, with graphical methods (sliding averages and polynomial adjustments), all with limited results, the previous information provided us with the idea of directly testing with the Upward Analytical Continuation (UAC) as a method of regional–residual separation of the relief or DEM. The elevation, in the calculation, was modified between 100 and 1000 m, with the result that the best solutions for the known onshore targets (and a few offshore) were obtained for the height of 500 m. In this way, the mathematical method selected for regional–residual separation in this publication has been that of the UAC at 500 m.

1.4 Conclusions

- The methods to evaluate the changes induced by the HCs in the superficial environment (on the HCs deposits), that is, morphometric, aeromagnetic detection and Induced Polarization, depend on the interaction of the microbes with the HCs that migrate vertically. In marine territories, the coincidence of all these anomalies, geomorphic and geophysical, allows to argue the combined use of unconventional methods of exploration, complementary to the seismic investigations.
- The interpretative procedure choose to obtain and evaluate the geomorphic anomalies related to the hardening of the surface rocks on the HCs deposits (determined from an assay-test method with known gas-oil targets data) has been the UCA at 500 m.

References

Donovan TJ, Friedman I, Gleason JD (1974) Recognition of petroleum-bearing traps by unusual isotopic compositions of carbonate-cemented surface rocks. Geology, pp 351–354

Donovan TJ, Noble RL, Friedman I, Gleason JD (1975) A possible petroleum-related geochemical anomaly in surface rocks, Boulder and Weld Counties. U.S. Geological Survey Open-File Report, Colorado, no 75–47, 11 p

Donovan TJ, Termain PA, Henry ME (1979) Late diagenetic indicators of buried oil and gas II direct detection experiment at cement and garza oil fields, Oklahoma and Texas, using enhanced Landsat I and II Images: U.S. Geological Survey Open-File Report, no 79–243, 45 p

Donovan TJ, Roberts A, Dalziel M (1981) Epigenetic zoning in surface and near surface rocks resulting from seepage induced redox gradients, Velma oil field, Oklahoma. Am Assoc Pet Geol Bull 65:919

Kartsev AA, Tabasaranskii ZA, Subbota MI and Mogilevskii GA (1959) Geochemical methods of prospecting and exploration for petroleum and natural gas (PA Witherspoon & WD Romey, Trans). Los Angeles, University of California Press, 349 p

Price LC (1985) A critical overview of and proposed working model for hydrocarbon micro seepage. Open-File Report 85–271. United States Department of the Interior Geological Survey. Denver, Colorado

Rosaire EE (1938) Shallow stratigraphic variations over Gulf Coast structures. Geophysics 3:96–121

Schumacher D (1996) Hydrocarbon-induced alteration of soils and sediments. In Schumacher D, Abrams MA (eds) Hydrocarbon migration and its near surface expression: AAPG memoir, vol 66, pp 71–89

Zeng H (1989) Estimation of the degree of polynomial fitted to gravity anomalies and its application. Geophys Prospect 37(8):959–973

Chapter 2
Use of the Digital Elevation Model for Offshore Hydrocarbons Exploration

Abstract It is exemplified, in the Block L of the Central East-North Cuba Sector (Cuba–Bahamas–Haiti Exclusive Economic Zone), the use of a Digital Elevation Model (DEM) at 90 × 90 m resolution for the detection of subtle positive geomorphic anomalies possibly related to hydrocarbon microseepage (vertical migration) on oil and gas targets. DEM data are integrated with geological, seismic, and Induced Polarization (IP) information. The results support the conclusion that DEM data provide a lower cost and fast offshore oil and gas preliminary exploration strategy. These data are useful serving to focus prospective areas to be checked with supplementary unconventional methods such as magnetic and IP and, after that, to propose the more expensive volumes for detailed 2D–3D seismic survey. In the particular case of Block L, the most intense geomorphic epicenter (geometric center of a geomorphic anomaly) does not coincide with a local dome of any Upper Jurassic horizons (presumably reservoirs), suggesting a possible link to a "Gas Chimney," related to a tectonic knot (local converging area of several tectonic dislocations) inside the North Cayo Fragoso structure. This hypothesis would also explain the observed intense IP anomaly. A "Gas Chimney" seems to be confirmed by 2D seismic information, for which the vertical zones of light hydrocarbon seepage are evident as well as their exit to the oceanic bed in the form of small podmarks.

Keywords Digital Elevation Model (DEM) · Hydrocarbon microseepage Geomorphic anomalies · Induced polarization · 2D seismic

2.1 Introduction

According to Schumacher (2014), the use of unconventional oil and gas exploration methods leads to a better assessment of prospective areas and exploration risks. In his work, a comparison of the results of the hydrocarbons non-conventional exploration (superficial geochemistry) with the results of the posterior drilling is carried out. This shows that in more than 2700 wells, all perforated after the end of the geochemical–geophysical (non-seismic) studies: Of perforated wells with positive results anomalies, 82% resulted in commercial discoveries, while only 11% of wells drilled

© The Author(s) 2018
M. E. Pardo Echarte et al., *Offshore Exploration of Oil and Gas in Cuba using Digital Elevation Models (DEMs)*, SpringerBriefs in Earth System Sciences, https://doi.org/10.1007/978-3-319-77155-7_2

without associated anomalies resulted in discoveries. Hence, the author proposes that drilling decisions should necessarily include a consideration of this type of studies, from which it turns out that the success rates of exploration increase significantly.

In the present chapter, the use of the DEM (90×90 m) for the detection of subtle positive geomorphic anomalies, possibly related to hydrocarbon microseepage (vertical migration) on offshore oil and gas targets, is exemplified in Block L of the Central East-North Cuba Sector (Cuba–Bahamas–Haiti Exclusive Economic Zone). It represents a contribution to the use of this type of anomalies in marine territories for the first time. These data are integrated with geological, seismic, and Induced Polarization (IP) information for the purposes of its validation and for the increase of the investigation's effectiveness.

For reasons of confidentiality, it is required to submit maps without coordinates.

2.2   Theoretical Framework

The Geological Premises that support the application of DEM and other unconventional geophysical methods in the detection of light hydrocarbons microseepage and their applications for offshore hydrocarbons exploration are the following:

• Light hydrocarbons microseepage (vertical migration) facilitates the formation of columns of mineralized rocks above oil and gas deposits (Price 1985; Pardo and Rodríguez 2016);
• The hydrocarbons in migration are oxidized by microbial action, creating a reducing environment (reducing chimneys) and producing, mainly, CO_2 and H_2S which drastically changes the pH/Eh of the system.
• Changes in pH/Eh result in changes in mineral stability, causing secondary mineralization of calcium carbonate and silicification, which result in denser, erosion resistant, and resistive surface materials (formation of positive geomorphic anomalies and maximum of resistivity).
• These changes in pH/Eh also precipitate magnetite/maghemite, and iron sulfides (pyrite, pyrrhotite, and greigite) or the coprecipitation of iron and/or manganese with calcite in the carbonate cements on the hydrocarbon deposits, which result in an increase in magnetic susceptibility and polarizability (which produce magnetic and IP anomalies).

The coincidence of all these anomalies, geomorphic and geophysical, allows us to argue the combined use of unconventional exploration methods, complementary to seismic investigations. Thus, it is proposed to carry out complementary investigations for offshore hydrocarbons exploration integrated by a complex of: DEM (90×90 m), magnetic and IP survey, the last ones in a mesh of profiles 2500×2500 m and observation step of 100 m. It should be noted that the expected positive geomorphic anomalies (residual maximum) vary in amplitude between 1 and 100 m, while magnetic, polarizability, and resistivity anomalies are also clearly observable.

The use of the DEM (90 × 90 m) provides a rapid way of preliminary offshore oil and gas exploration, being useful to focus the areas to be raised with complementary non-conventional methods (Magnetic and IP) whose complex results would argue the volumes for the more expensive detail 2D–3D seismic. This is the main contribution of this chapter.

2.2.1 Definitions

Digital Elevation Model: Understand as the grid (90 × 90 m) in GeoSoft format (minimum curvature interpolation), obtained from the satellite radar and bathymetric database of the relief, whose source is http://www.cgiar-csi.org/data/srtm-90m-digital-elevation+ETOPO2.

Geomorphic anomaly: It is understood as the residual relief result of subtracting to the DEM a regional tendency.

Tectonic knot: Understand as the local confluence zone of several tectonic dislocations.

Geomorphic epicenter: Understand as the geometric center of a geomorphic anomaly.

2.2.2 Sources of Information and Processing

The DEM (90 × 90 m) used in this work was taken from Sánchez Cruz et al. (2015) with source at http://www.cgiar-csi.org/data/srtm-90m-digital-elevation+ETOPO2. The processing of these data consisted in the regional–residual separation of the relief (allowing emphasizing lower amplitude anomalies), from the calculation of the Upward Analytical Continuation (UAC) to 500 m, subtracting this from the observed relief (DEM).

Seismic data were taken from Zobnina et al. (2010). The processing for the obtaining the deep migrated surfaces for the different Upper Jurassic horizons was using the Petrel software version 2015.

IP data were taken from Garina et al. (2010).

The automated processing and interpretation of the georeferenced geophysical–geological information were carried out with the Oasis Montaj (R) System, version 7.01.

2.3   Experimental Development

The geographical location for the blocks of the Cuba–Bahamas–Haiti Exclusive Economic Zone (Central East-North Cuba Sector) is indicated in Fig. 2.1 (López Rivera et al. 2012).

The information on the expected geological model for the two main Prospects in this area (North Cayo Fragoso and Cayo Guillermo) (López Rivera 2012), states that the gas and oil objectives for the North Cayo Fragoso Prospect are related to the Upper Cretaceous and the Upper Jurassic.

Due to the results of the 2D seismic campaigns carried out in Block L, the North Cayo Fragoso Prospect was established in the southwestern region of the block, where the exploratory well L01X was drilled (González Rodríguez et al. 2014). This well reached a depth of 1803.9 m, crossing the Upper Cretaceous layers without finding the expected target and being interrupted by failures before reaching the top of the Jurassic.

According to the results of the IP works (Garina et al. 2010), only three objects of the category plus perspective (in yellow) were identified. These are limited to the narrow and elongated structure identified in the DEM, complicated by faults, associated with the upper part of the Cretaceous and Jurassic sediments in the southwestern region of Block L (North Cayo Fragoso Prospect) (Fig. 2.2). According to the authors of the report, prospective objects could be associated with the presence of hydrocarbons in both Jurassic and Cretaceous sediments (the latter discarded by well L01X); the separation of the IP response from the Jurassic and Cretaceous reservoirs would not be possible.

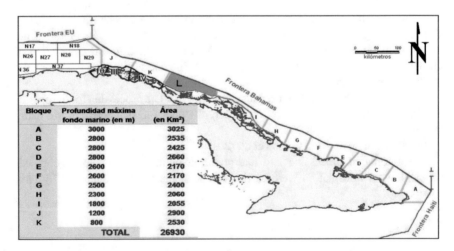

Fig. 2.1 Blocks from the Cuba–Bahamas–Haiti Exclusive Economic Zone (Central East-North Cuba Sector) Taken from López Rivera et al. (2012)

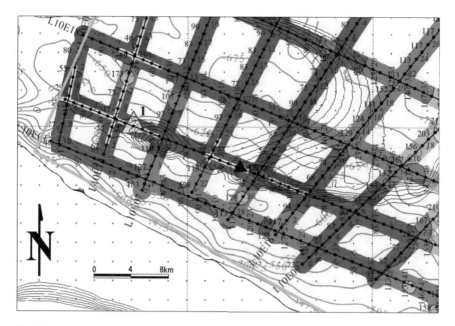

Fig. 2.2 Induced polarization results versus contours of the residual DEM at 500 m. In yellow, anomalous IP intervals of the first category. The yellow triangle corresponds to the exploratory well recommended to be drilled in Block L by the results of IP. Taken from Garina et al. (2010), modified by M. Pardo

In addition, in the opinion of these authors, the identified anomalous objects are located, laterally, near the axial parts of the structure, but partially do not coincide with them. Taking into account the complex geological conditions and the use of 2D seismic data, they assume the presence of some inaccuracies in the structural plane. On the other hand, the carbonate reservoir could be characterized by spatial heterogeneity. In this case, likely traps that generally tend to the axis of the structure can be displaced from the plane of the local arcs. Thus, in order to determine the relationship of IP anomalies with supposed accumulations of hydrocarbons in carbonate reservoirs of Jurassic–Cretaceous sediments, the authors of the IP survey proposed the drilling (not executed) of an exploratory well at the point SL 41 (most intense anomaly) of the L10E02 line, indicated by a yellow triangle in the results map (Fig. 2.2).

2.4 Results

In the particular case of Block L, the most interesting results are centered on the North Cayo Fragoso Prospect, where the exploratory well L01X was drilled. This well reached a depth of 1803.9 m, crossing the Upper Cretaceous layers without

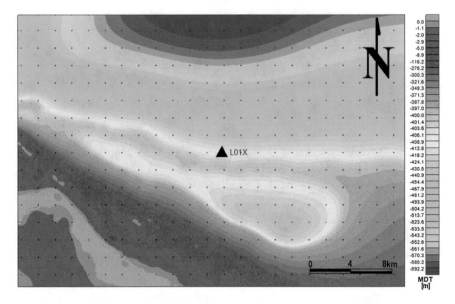

Fig. 2.3 DEM of the North Cayo Fragoso Prospect area, Block L

finding the expected target and being interrupted by failures before reaching the top of the Jurassic.

The DEM of the North Cayo Fragoso Prospect area is presented in Fig. 2.3. In the same one, it appears the location of the drilled exploratory well L01X (black triangle). The corresponding residual DEM at 500 m is presented in Fig. 2.4, where the contours of −1.0, −0.5, 0, 0.5, 1.0, 1.5, 2.0, and 2.5 m, respectively, have been represented. According to them, a long, narrow, sub-latitudinal geomorphic anomaly is established, corresponding to the Prospect's limits in good approximation. The anomaly has a more intense epicenter to the west, which maintains a good spatial correspondence with the main IP anomaly of the area (the most intense yellow triangle) and the exploratory well recommended by the authors of the IP survey (Fig. 2.2). In Fig. 2.2, it is also observed that the location of the exploratory well L01X (black triangle) was successful, from the point of view that it corresponds with an anomalous seismic-geomorphic-IP complex.

2.5 Discussion

An examination of the spatial correspondence between the geomorphic anomaly (contours) and the different Upper Jurassic horizons generated from the seismic interpretation with the software Petrel from Schlumberger (Figs. 2.5, 2.6, and 2.7) reveals a satisfactory overall coincidence, but no preference can be established for any

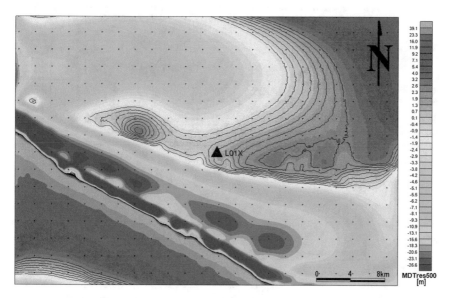

Fig. 2.4 Residual DEM at 500 m (DEMres500) from the North Cayo Fragoso Prospect area, Block L

specific one. The fact that any local dome of one of the Upper Jurassic horizons does not reflect the most intense geomorphic epicenter suggests a possible relationship with a "Gas Chimney" linked to a tectonic knot within the structure, which would also explain the intense observed anomaly of IP. A "Gas Chimney" seems to be confirmed by the 2D seismic information, shown in Fig. 2.8, which shows the section for the attribute "Intensity of Reflection" corresponding to a transverse line (yellow, coarse color) that intercepts the geomorphic anomaly by its center. According to this figure, the vertical zones of light hydrocarbons seepage and their exit to the oceanic bed in the form of small podmarks become evident. In any case, the anomalous geomorphic-IP complex of the West part of the North Cayo Fragoso Prospect reflects the location of the most intense hydrocarbon microseepage processes taking place in this Prospect and, apparently, throughout Block L.

2.6 Conclusions

- The use of the DEM provides a cost-effective and expeditious approach to research for offshore oil and gas exploration, which may be useful in targeting the areas to be raised with complementary non-conventional (magnetic and IP) methods whose results complex would argue the more expensive volumes of detail 2D–3D seismic.

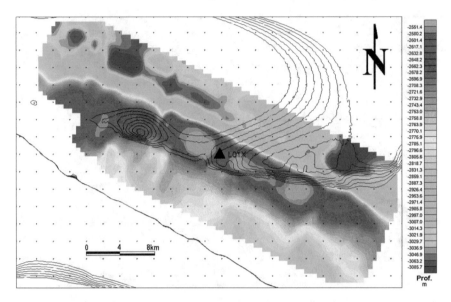

Fig. 2.5 Oxfordian surface versus contours of the residual DEM at 500 m

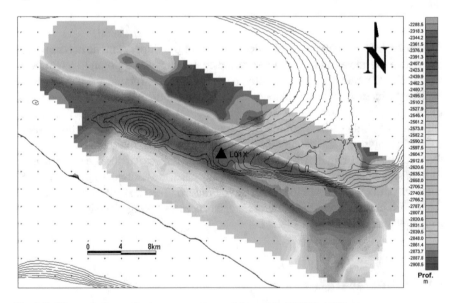

Fig. 2.6 Kimmerigdian surface versus contours of the residual DEM at 500 m

- In the particular case of Block L, the results confirm the greater interest in the North Cayo Fragoso Prospect, where the exploratory well L01X was drilled, although this well has already crossed the Upper Cretaceous layers without finding the expected objective.

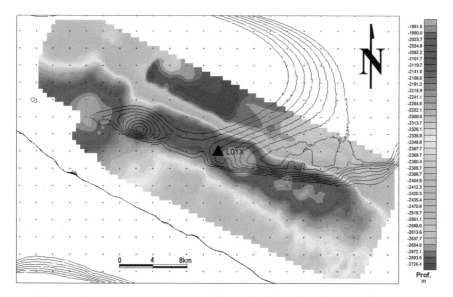

Fig. 2.7 Tithonian surface versus contours of the residual DEM at 500 m

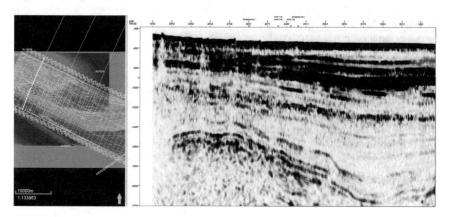

Fig. 2.8 2D seismic data versus geomorphic anomaly

- The examination of the spatial correspondence between the positive geomorphic anomaly (contours) and the different Upper Jurassic horizons reveals a satisfactory overall coincidence. The fact that any local dome of one of the Upper Jurassic horizons does not reflect the most intense geomorphic epicenter suggests a possible relationship with a "Gas Chimney," linked to a tectonic knot within the structure, which would also explain the intense observed anomaly of IP. A "Gas Chimney" seems to be confirmed by the 2D seismic information, for which the vertical zones of light hydrocarbon seepage are evident, as well as, their exit to the oceanic bed in the form of small podmarks. In any case, the anomalous geomorphic-IP complex of the West part of the North Cayo Fragoso Prospect is related to the more intense hydrocarbon microseepage processes taking place in this prospect and, apparently, throughout Block L.

References

ETOPO 2. Fuente: http://worldgrids.org/doku.php/wiki:dem_sources

Garina SJ, Ivanov SA, Pesterev IJ, et al (2010) Report on the object: "Carrying out offshore oil-and-gas geophysical prospecting by differentially-normalized method of electro investigation (DNME) on Block L of Republic of Cuba". Informe Técnico, (Inédito). CEINPET, La Habana, 92 p

González Rodríguez EM, Prieto F, Suárez V, et al (2014) Proyecto 7075 "La Exploración de Hidrocarburos Costa Afuera (Aguas Profundas): Sectores Centro Oriental y Cuba Sur" Informe del Grado de Estudio Geólogo – Geofísico en la zona de aguas profundas de Cuba Centro – Oriental (costa norte). Informe Técnico, (Inédito). CEINPET, La Habana, 75 p

López Rivera JG (2012) Petroleum systems and exploratory models in blocks 17–18 & L. Presentación en power point. CEINPET, La Habana, 114 d

López Rivera JG, Domínguez BR, Morales I, Delgado O, et al (2012) Informe "Apertura a la exploración petrolera del sector Cuba – Bahamas – Haití de la zona económica exclusiva de la República de Cuba". "Play PRESAL". Informe Técnico, (Inédito). CEINPET, La Habana, 43 p

Pardo Echarte ME, Rodríguez Morán O (2016) Unconventional methods for oil & gas exploration in Cuba. Springer Briefs Earth Syst Sci. https://doi.org/10.1007/978-3-319-28017-2

Price LC (1985) A critical overview of and proposed working model for hydrocarbon microseepage. US Department of the Interior Geological Survey. Open-File Report 85–271

Sánchez Cruz R, Mondelo F, et al (2015) Mapas Morfométricos de la República de Cuba para las Escalas 1: 1,000,000–50,000 como apoyo a la Interpretación Geofísica. Memorias VI Convención Cubana de Ciencias de la Tierra, VIII Congreso Cubano de Geofísica. http://www.cgiar-csi.org/data/srtm-90m-digital-elevation+ETOPO2

Schumacher D (2014) Minimizing exploration risk: the impact of hydrocarbon detection surveys for distinguishing traps with hydrocarbons from uncharged traps. GeoConvention 2014: FOCUS

Zobnina NI et al (2010) Informe de las campañas sísmicas 2D realizadas en el Bloque L. Informe Técnico, (Inédito). CEINPET, La Habana, 57 p

Chapter 3
Use and Validation of the Digital Elevation Model for Offshore Hydrocarbon Exploration in the Region of Habana–Matanzas, Cuba

Abstract The use of the DEM to a sampling of 90 × 90 m for the detection of positive residual geomorphic anomalies possibly related to hydrocarbon microseepage on known oil and gas fields or production areas is validated in the North Cuban Oil Strip (NCOS) of the Habana–Matanzas region, Cuba. The data of the residual DEM at 500 m are integrated with the information of the exploitation wells and the mapped seismic structures. Sixty percent of the known offshore oil fields or production areas of the NCOS have positive residual anomalies; of them, 55% have amplitudes between 10 and 25 m and 22% are of amplitude >50 m. All fields in operation are characterized by seismic structures. Seventy-nine percent of the mapped seismic structures intercept positive residual geomorphic anomalies, of which 53% have amplitudes between 10 and 25 m and 27% are >50 m. Mapped seismic structures (and deposits) that do not correspond to positive residual anomalies are located within a strip of increased positive values observed from land up to 1–2 km offshore in the Mariel–Matanzas Bay sector, where, because of the character of the residual DEM, it is impossible to detect isolated anomalies within it. The results allow establishing that DEMs provide a rapid form of preliminary offshore oil and gas exploration.

Keywords Digital Elevation Model (DEM) · Hydrocarbon microseepage Geomorphic anomalies · 2D–3D seismic · Oil exploitation drilling · North cuban oil strip

3.1 Introduction

In Chap. 2, the use of the DEM (90 × 90 m) for the detection of subtle positive residual geomorphic anomalies possibly related to hydrocarbon microseepage on offshore oil and gas targets is exemplified in Block L (Cuba–Bahamas–Haiti Exclusive Economic Zone, Central East-North Cuba sector), representing, for the first time, the use of such anomalies for exploratory purposes in marine territories.

The objective of this chapter is to validate, more extensively, this result in a known gasopetroliferous region. In this way, the DEM (90 × 90 m) data are integrated with the information of exploitation wells and seismic of the north coast of Habana—Matanzas region, comprised between Morro–Cabañas and Varadero–Cárdenas, also known as North Cuban Oil Strip (NCOS) .

According to the collective of authors (2009), the NCOS covers the coastal belt of Habana and Matanzas provinces, including the adjacent marine platform, about 5 km wide and 150 km long, where most and the most large deposits of the country have been discovered, although it is possible that it extends even more toward the west and the east.

In this strip, the two largest deposits of Cuba (Varadero and Boca de Jaruco) were discovered at the end of the 1960s', although their current extension was established from the 1990s', in which with the participation of foreign companies were carried out several campaigns of 2D and 3D seismic. These revealed a train of structures along the coast where exploratory drilling confirmed the deposits Puerto Escondido–Canasí, Yumurí–Seboruco and the extension to the west of the Varadero deposit (Varadero Oeste).

Subsequently, other deposits have been discovered, to mention some: Santa Cruz, Tarará, Bacuranao, Jibacoa, Habana del Este, and Morro–Cabañas. The density of the oil found fluctuates between 11° and 14° API, achieving, from the directed drilling with large angles (horizontal), wells with stable entries of the order of up to 4000 barrels/day.

The subsurface geology of these sectors is one of the best known in the country. According to the aforementioned source, it is characterized by several levels of ramp folds against inverse faults of North American Continental Margin rocks and their coverage. These folds have probably been further complicated by shear accidents. The deformed rocks cover an interval of age that goes from the Jurassic to the Eocene. The reservoirs are represented by heavily fractured and leached limestones covered by a seal of clay from the Paleocene to the Eocene; they have their analogues in the outcrops of rocks of the North American Continental Margin in Central Cuba.

The stacks of several ramp anticline folds are the main exploratory objective. These form antiforms that are mapped with great difficulty by seismic. The poor image is the main obstacle to the development of the exploratory works; only it is observed as horizon with dynamic high definition the one directly related to the envelope of the scales folds. Hence, it is so important to have other complementary exploratory tools that allow a better evaluation of the prospective areas, with an increase in the geological effectiveness of the exploration and a reduction of its risks. In this sense, the use of the DEM (90 × 90 m) offers an interesting and novel contribution.

For reasons of confidentiality, maps are presented without coordinates and the reference to oil production is made only in very general terms.

3.2 Theoretical Framework

As seen in the two previous chapters, the Geological Premises that support the application of DEM and other non-conventional geophysical methods in the detection of light hydrocarbon microseepage and its applications for the offshore hydrocarbon exploration are the following:

- Light hydrocarbon microseepage (vertical migration) facilitates the formation of columns of mineralized rocks above oil and gas deposits (Price 1985; Pardo and Rodríguez 2016).
- The hydrocarbons in migration are oxidized by microbial action, creating a reducing environment (reducing chimneys) and producing, mainly, CO_2 and H_2S, which drastically changes the pH/Eh of the system.
- Changes in pH/Eh result in changes in mineral stability, causing secondary mineralization of calcium carbonate and silicification, which result in denser, erosion-resistant, and resistive surface materials (formation of positive geomorphic anomalies and maximum of resistivity).
- These changes in pH/Eh also promotes precipitates of magnetite/maghemite, and iron sulfides (pyrite, pyrrhotite, and greigite) or the coprecipitation of iron and/or manganese with calcite in the carbonate cements on the hydrocarbon deposits, which result in an increase in magnetic susceptibility and polarizability (which produce magnetic and IP anomalies).

The coincidence of all these anomalies, geomorphic and geophysical, allows arguing the combined use of unconventional exploration methods, complementary to seismic investigations. Thus, the use of the DEM (90 × 90 m) provides a rapid way of preliminary offshore oil and gas exploration, being useful to focus the areas to be raised with complementary non-conventional methods (magnetic and IP) whose complex results would argue the volumes for the more expensive detail 2D–3D seismic.

3.3 Materials and Methods

3.3.1 Sources of Information and Processing

The DEM (90 × 90 m) used in this work was taken from Sánchez Cruz et al. (2015), with source at: http://www.cgiar-csi.org/data/srtm-90m-digital-elevation+ETOPO2. The processing of these data consisted of the regional–residual separation of the relief from the calculation of the Upward Analytical Continuation (UAC) to 500 m, subtracting this from the observed relief (DEM), thus allowing emphasizing residual anomalies of less vertical amplitude.

The data from the drilled exploitation wells were taken from the Ceinpet—Exploration Information Unique System (EIUS) Database (Pico Peña 2012).

The production information is the result of personal communication with the specialists of the Extraction and Oil Drilling West Company (EODC), Jesús Ríos Montano (Exploration Geologist) and Víctor Moya (Head of the Department of Reservoir Engineering).

The information of the seismic structures was taken from the exploration reports of Domínguez et al. (2016) and González et al. (2015).

The processing of the information was done using the Petrel software, version 2015, from Schlumberger.

Part of the automated processing and interpretation of the georeferenced geophysical–geological information, related to the morphotectonic regionalization, was carried out with the Oasis Montaj (R) System, version 7.01.

The morphotectonic regionalization was performed, for the Mariel–Cárdenas territory, from the map of the residual DEM at 500 m by tracing morphotectonic alignments following the linearity, gradients, inflections, and interruption of the contours, which separate different tectono-structural elements characterized by different levels of amplitude of positive residual anomalies.

The comparative analysis between the morphotectonic scheme with positive residual geomorphic anomalies versus the known oil fields and/or production areas and versus the mapped seismic structures was performed according to the following criteria:

1. Number of known oil fields and/or production areas corresponding to positive residual geomorphic anomalies. From the total, there were separated those corresponding to the amplitude levels [−20 to 10 m]; [10 to 25 m]; [25 to 50 m]; and [>50 m], also characterizing the different levels of production in each known area or oil field.
2. Number of known oil fields and/or production areas that do not correspond to positive residual geomorphic anomalies.
3. Number of mapped seismic structures corresponding to positive residual geomorphic anomalies. From the total, there were separated those corresponding to the amplitude levels [−20 to 10 m]; [10 to 25 m]; [25 to 50 m]; and [>50 m].
4. Number of mapped seismic structures that do not correspond to positive residual geomorphic anomalies.

3.4 Results

3.4.1 Morphotectonic Regionalization

A limiting characteristic of the residual DEM at 500 m (Fig. 3.1) is that between Mariel and the Bay of Matanzas there is a strip of positive values observed from

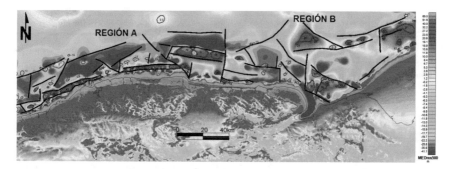

Fig. 3.1 Morphotectonic scheme with positive residual geomorphic anomalies from the residual DEM at 500 m

land up to 1–2 km offshore, due to the character of the marine platform in this sector, which prevents detection of possible anomalies (structures) isolated within the same.

From the morphotectonic point of view, in the Mariel–Cárdenas territory two regions (A and B) can be separated, with different characters in the distribution of the tectono-structural elements and amplitude of the positive residual anomalies, from an alignment NS to the level of Playa Jibacoa (Fig. 3.1). The region A (Mariel—Playa Jibacoa) presents alignments and chains (trains) of anomalies with amplitudes between 25 and 70 m (predominating those >50 m) parallel to the coast, mostly corresponding to non-drilled structures. As a rule, region B (Playa Jibacoa–Cárdenas) presents independent blocks arranged in a chaotic form with anomalies of amplitudes between −18 and 25 m, predominating those of 10–25 m, being in this region the fields of greater productivity of the NCOS.

It should be clarified that the residual anomalies indicated with negative sign are also positive, only within a background of negative values.

In region A, in general, all anomalies are located parallel to the coast. It stands out from the west to east three chains (trains): Mariel–Habana with a distance of 2–4 km from the coast (amplitudes 25–70 m); Playas del Este–Boca de Jaruco with a distance of 3–4 km from the coast (amplitudes 25–70 m); and Boca de Jaruco–Playa Jibacoa with a distance of 10–12 km (amplitudes 60–110 m). Below, to the south of the first mentioned chain, between Mariel and Jaimanitas, at a distance of 1.5 km, a series of small anomalies are arranged with amplitudes between 15 and 45 m. Above, north of the first two chains of anomalies, between Mariel and Boca de Jaruco, at a distance of 6–8 km, another series of anomalies with amplitudes between −15 and 20 m are arranged. Further north, between Mariel and Jaimanitas, at a distance of 14–16 km, another series of anomalies with amplitudes between −15 and 15 m are arranged. Below, to the south of the third chain mentioned, between Boca de Jaruco and Playa Jibacoa, there are three anomalies isolated at a distance between 3 and 5 km, with amplitudes between 15 and 30 m.

Table 3.1 Known oil fields and/or production areas corresponding to positive residual geomorphic anomalies

Oil field	Anomalies amplitude (m)
Yumurí–Seboruco	−20 to 10
Santa Cruz	10–25
Bacuranao	10–25
Varadero Oeste	10–25
Canasí	10–25
Tarará	25–50
Cabaña	10–25
Habana del Este	>50
Seboruco Este	>50

In the region B, parallel to the coast, between Playa Jibacoa and Bacunayagua, at a distance of 2–2.5 km, three anomalies are arranged with amplitudes between 15 and 25 m. In the so-called Seboruco Block, parallel to the coast at a distance of 2.5 km, a series of anomalies are arranged with amplitudes between −18 and −10 m. Further north, within this same block are located two other anomalies of −13 and −10 m of amplitude, located at a distance of 5 and 8 km, respectively. In the so-called Varadero Block, between Matanzas Bay and the Hicacos Peninsula, a series of anomalies are located on a SE-NO course, at a distance of 5–9 km, which includes at its eastern end (to 5 km) the Varadero Oeste oil field, with amplitudes between 15 and 25 m. Within this same block, to the south, at a distance of 3–5 km are located three anomalies with amplitudes between 20 and 80 m. Further north, between the bay of Matanzas and Cárdenas, at a distance of 20–30 km, a chain of anomalies with a SO-NE course is arranged, with amplitudes between 15 and 25 m. To the south of it, and to the north of the bay of Cárdenas, with the same course, are located four anomalies with equal amplitudes.

3.4.2   Comparative Analysis Between the Morphotectonic Scheme with Positive Residual Geomorphic Anomalies Versus the Known Oil Fields and/or Production Areas

All anomalies associated with fields in operation have been reached by, at least, one field well. It is noteworthy that the most productive fields at present have anomalies that do not exceed amplitudes of 25 m, as shown in Table 3.1.

Sixty percent of the known offshore oil fields or production areas of the NCOS have positive residual anomalies; of them, 55% have amplitudes between 10 and 25 m and 22% are of amplitude >50 m. In Fig. 3.2, the producing wells that intercept the referred anomalies are represented.

Due to the nature of the marine platform in the Mariel–Matanzas Bay sector, where it is impossible to detect isolated anomalies (structures), the remaining 40%

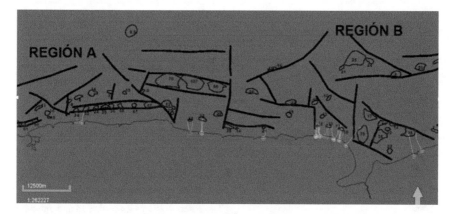

Fig. 3.2 Producing wells that intercept the positive residual anomalies associated with different deposits or production areas known in the NCOS, offshore

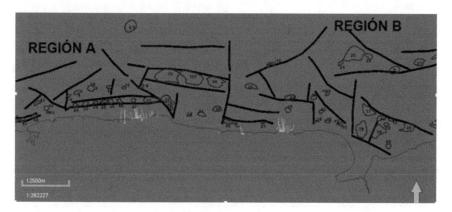

Fig. 3.3 Producing wells associated with different deposits or production areas known in the NCOS, offshore, that do not intercept positive residual anomalies

of the known deposits or areas of production of the NCOS, offshore, are located. These producing areas are: Cojímar–Alamar, Guanabo, Vía Blanca, Jibacoa, Boca de Jaruco, and Puerto Escondido. Figure 3.3 shows the producing wells that do not intercept positive residual geomorphic anomalies.

A commentary on the production of petroleum in known oil fields, or areas of production, coincident and not coincident with positive residual geomorphic anomalies is given in Tables 3.2 and 3.3, respectively.

Table 3.2 Known oil fields or production areas coinciding with positive residual geomorphic anomalies and criteria about its production

Oil field	Criteria about its production
Yumurí–Seboruco	The wells that intercept geomorphic anomalies are development wells with good production
Santa Cruz	Producer of oil
Bacuranao	Bacuranao 300 well currently drilling but with high prospects
Varadero Oeste	Excellent oil producer
Canasí	The Canasí 9 well that intercepts the geomorphic anomaly had good production but is in the declining phase
Tarará	The Tarara N201 well was drilled and did not reach the target, remaining in the seal, but with manifestations of heavy oil
Cabaña	The well Cabaña 100 had a test of production with good results and discovers oil. The well is liquidated by collapse
Habana del Este	The Habana 102 well had a good production of more than 100 m^3/d. It is liquidated by the high concentration of hydrogen sulfide in the gas
Seboruco Este	Good oil producer

Table 3.3 Known oil fields or production areas not coinciding with positive residual geomorphic anomalies and criteria about its production

Oil field	Criteria about its production
Cojímar–Alamar	There is a producing well (Cojímar100)
Puerto Escondido	Producer of oil
Guanabo	Producer of oil
Vía Blanca	Producer of oil
Jibacoa	Producer of gas
Boca de Jaruco	Producer of oil and gas

3.4.3   Comparative Analysis Between the Morphotectonic Scheme with Positive Residual Geomorphic Anomalies Versus the Mapped Seismic Structures

All fields under exploitation are characterized by seismic structures. It should be noted that the structures currently mapped have, in the main, anomalies that do not exceed amplitudes of 25 m, as shown in Table 3.4.

Seismic structures interpreted in the study region were analyzed, with 79% of the mapped structures intercepting positive residual geomorphic anomalies, of which 53% have amplitudes between 10 and 25 m and 27% are >50 m. Figure 3.4 shows

Table 3.4 Mapped seismic
structures that correspond to
positive residual geomorphic
anomalies

Structure	Anomalies amplitude (m)
Varadero Oeste	10–25
Yumurí–Seboruco	−20 to 10
Yumurí Norte	−20 to 10
Puerto Escondido–Norte	10–25
Fraile Norte	10–25
Santa Cruz	10–25
Boca de Jaruco–Vía Blanca	>50
Guanabo Norte	>50
Bacuranao–Tarará	25–50
Bacuranao–Tarará Norte	10–25
Habana del Este–Cabañas	>50
Cojímar–Alamar	10–25
Varadero Oeste–Sur	10–25
Seboruco Este	>50
Seboruco Este–Norte	10–25

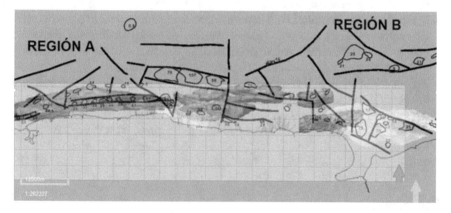

Fig. 3.4 Mapping seismic structures that intercept positive residual anomalies in the NCOS, off-shore

the mapped seismic structures that intercept positive residual anomalies in the NCOS, offshore.

Mapped seismic structures that do not correspond to positive residual anomalies are located within the strip of increased positive values observed from land up to 1–2 km offshore (due to the character of the marine platform in the Mariel–Matanzas Bay sector), where it is impossible to detect the anomalies (structures) isolated inside it. These structures, which represent 21%, are Guanabo–Vía Blanca, Jibacoa, Boca de Jaruco, and Puerto Escondido.

3.5   Conclusions

- From the morphotectonic point of view, in the Mariel–Cárdenas territory two regions (A and B) can be separated with different characters in the distribution of the tectono-structural elements and amplitude of the positive residual anomalies, from an alignment with course NS to the level of Playa Jibacoa. The region A (Mariel–Playa Jibacoa) has alignments and chains (trains) of anomalies with amplitudes between 25 and 70 m (predominant >50 m) parallel to the coast, mostly corresponding to non-drilled structures. As a rule, region B (Playa Jibacoa–Cárdenas) presents independent blocks arranged in a chaotic form with anomalies of amplitudes between −18 and 25 m, predominating those of 10–25 m, being in this region the fields of greater productivity of the NCOS.
- Sixty percent of the known offshore oil fields or production areas of the NCOS have positive residual anomalies; of them, 55% have amplitudes between 10 and 25 m and 22% are of amplitude >50 m.
- All fields in operation are characterized by seismic structures. Seventy-nine percent of the mapped seismic structures intercept positive residual geomorphic anomalies, of which 53% have amplitudes between 10 and 25 m and 27% are >50 m.
- Mapped seismic structures (and deposits) that do not correspond to positive residual anomalies are located within a strip of increased positive values observed from land to 1–2 km offshore in the Mariel–Matanzas Bay sector, where it is impossible to detect the anomalies (structures) isolated within the same.
- The results allow establishing that the DEM provides a rapid form of preliminary offshore oil and gas exploration. Also is useful for focusing the areas to be raised with complementary unconventional methods such as magnetic and IP surveys and to propose the most expensive volumes to be raised with the detail 2D–3D seismic.

References

Colectivo de Autores (2009) Expediente Único del Proyecto 6004, "Exploración en la Franja Norte de Crudos Pesados". Archivo, Centro de Investigaciones del Petróleo (Ceinpet), La Habana, Cuba

Domínguez BR, González EM, et al (2016) Resultados de la exploración sísmica 3D y caracterización de prospectos exploratorios para petróleo y gas en los Bloques 7 y 7A, al Norte de La Habana y Mayabeque-Proyectos I + D 7080 y 9015. Ceinpet, La Habana, Cuba

ETOPO 2. Fuente: http://worldgrids.org/doku.php/wiki:dem_sources

González EM, Reyes O, et al (2015) Informe Parcial sobre la actualización del modelo geológico Varadero Oeste, Bahías de Matanzas y Cárdenas -Proyectos I + D 7080 y 9000. Ceinpet, La Habana, Cuba

Pardo Echarte ME, Rodríguez Morán O (2016) Unconventional Methods for Oil & Gas Exploration in Cuba. Springer Briefs Earth Syst Sci. https://doi.org/10.1007/978-3-319-28017-2

Pico Peña R (2012) Sistema Único para la Información de la Exploración (SUIE)-Proyecto I + D 7038. Ceinpet, La Habana, Cuba

Price LC (1985) A critical overview of and proposed working model for hydrocarbon microseepage. US Department of the Interior Geological Survey. Open-File Report 85–271

Sánchez Cruz R, Mondelo F, et al (2015) Mapas Morfométricos de la República de Cuba para las escalas 1: 1,000,000–1: 50,000 como apoyo a la Interpretación Geofísica. Memorias VI Convención Cubana de Ciencias de la Tierra, VIII Congreso Cubano de Geofísica. Fuente: http://www.cgiar-csi.org/data/srtm-90m-digital-elevation+ETOPO2

Chapter 4
Use of Seismic Attributes and Geomorphic Anomalies for the Evaluation of Prospective Zones in the Cuban North Oil Strip

Abstract The search for new areas for the production of hydrocarbons is a constant challenge in oil industry. The geological intricacies inherent in folded and thrusted belts favor the increase in the degree of error in oil exploration of the north of western Cuba. With the objective of revealing new areas of petroleum interest, a study was carried out in regions of the Northern Cuban Thrusted and Folded Belt (NCTFB) where there are known oil producing fields in the North Cuban Oil Strip (NCOS), helping to establish analogies in unknown areas. Positive residual geomorphic anomalies, obtained from residual at 500 m Digital Elevation Model (DEM) data (possibly related to hydrocarbon microseepage on offshore oil and gas targets), were used. Also were used the seismic attributes "Relative Acoustic Impedance" (RAI) and "Genetic Inversion" (GI), from which volumes of assisted properties were generated with data of wells (responding to the presence of reservoirs). The analysis of the coincidence of the geomorphic anomalies with the already drilled structures allows proposing the existence of a relation between its amplitude and the depth of the structures, noting a decrease of this with the depth of the drilled objectives. The response of the seismic attribute in question, linked to the presence of geomorphic anomalies in non-drilled zones, allows to define the possible extension of reservoirs in operation and new areas of interest not yet drilled.

Keywords Offshore oil and gas exploration · Thrusted and folded belts
Seismic structures · Geomorphic anomalies · Seismic attributes
Reservoirs of the North Cuban Oil Strip

4.1 Introduction

In the measure that the number of wells directed to the producing structures in the gasopetroliferous northern Cuban region (NCTFB) has been increasing, it has been demonstrated the structural complexities that govern the formation of the traps. The presence of elements that complicate the geometry of the tectonic stacks, characteristic of the process of forming thrusted and folded belts, makes it difficult to detect new prospects for the production of hydrocarbons.

© The Author(s) 2018   29
M. E. Pardo Echarte et al., *Offshore Exploration of Oil and Gas in Cuba
using Digital Elevation Models (DEMs)*, SpringerBriefs in Earth System Sciences,
https://doi.org/10.1007/978-3-319-77155-7_4

Petroleum deposits in the marine sector of bathymetrically transitional waters belonging to the NCOS are structurally formed by overcrowded, preferential SW–NE scales of fractured carbonate rocks of the North American Continental Margin. They are arranged in the form of triangles or tectonic stacks, which constitute the reservoir, whose regional seal is the Vega Alta clay formation. The scales, because of the overlap occurred in the critical stage of the Eocene, found at their top the area with better reservoir properties.

The search for new areas of interest in developing fields requires the use of resolution methods such as seismic, with the objective of evaluating fluid saturations in the reservoir, for which time lapses techniques are conventionally applied (seismic 4D). In the absence of this advanced method and in view of the need to find new prospective areas for the production of hydrocarbons, seismic attributes are used (Vargas-Meleza et al. 2004), together with information from wells, facilitating the detection of anomalies associated with the presence of fluids. The use of other complementary methods such as the Digital Elevation Models (DEMs) allows the detection of positive residual geomorphic anomalies possibly linked to the hydrocarbons microseepage, essentially vertical, on offshore oil and gas targets, making it very useful for the planimetric location of new prospects to be evaluated with seismic attributes. Consequently, the chapter aims to test, for the first time, a prospective interpretation methodology that combines information on the location of possible production structures, based on positive residual geomorphic anomalies, with the corresponding evaluation of the possible saturation of fluids, from the use of seismic attributes. Its validity will be given by the practical confirmation of the recommended prospective areas for hydrocarbons.

For reasons of confidentiality, the figures are presented without coordinates.

4.2   Theoretical Frame

The spatial coincidence of the anomalies, geomorphic and geophysical, allows us to argue the combined use of unconventional exploration methods complementary to seismic investigations. Seismic allows the delimitation of structures that may or may not be saturated with hydrocarbons, so the establishment of a link of the same to the well data through the GI attribute allows us to obtain fluid saturation criteria in the mentioned structures, supported by the criteria for the existence of hydrocarbon microseepage processes provided by the DEM.

The characterization of oil deposits from seismic data under reservoir conditions implies the understanding of phenomena associated with the propagation of seismic waves and their relationship with reservoir properties, as well as the temporal and spatial variations of these properties. The combined use of geophysical well logs and seismic attributes, assisted by predictive mathematical techniques, could be an effective method in the characterization of carbonate reservoirs, as proposed in Pedro-Pompa, Reyes-Paredes, Veiga-Bravo y Miró-Pagés (in press).

Usually, for a drilled well there are measurements that give a correct idea of the physical and elastic properties (velocity, density, porosity, etc.) of the rocks in their vicinity. However, to understand the behavior of these properties far from the well we have to stick to the data of the seismic volumes. Current studies of rock physics allow us to relate these properties to the seismic data and to infer their variations in a lateral and vertical sense. Thus, by studying the physical properties of the rocks of the hydrocarbon reservoirs in the wells and using the seismic information, new prospects can be delineated away from the wells (Ekwe et al. 2012).

The attributes are mathematical transformations that are made to the seismic data with the objective of highlighting specific characteristics of interest for a better visualization. Your choice depends on the set of data you have, and what you want to highlight. These were first introduced in the early 1970s, and they are now widely used for the prediction of lithological and petrophysical properties of reservoirs (Chopra and Marfurt 2006).

Carbonates have been shown to present many challenges, such as lateral inconsistency of rock properties, subtle facies changes, shortened seismic expression (due to their high speeds), and dispersion velocities (due to heterogeneity of permeability) (Wang and Dopkin 2012). However, a good selection of attributes can give positive results in their recognition.

Many may be the attributes to be used; however, it has been shown that, for carbonated rocks, the following are useful:

- **Relative Acoustic Impedance**. It reflects the contrast of physical properties and provides a better fit to the well domain. This physical attribute assumes zero-phase data. It is generated by the integration of the seismic trace and a subsequent high-pass filter to remove arbitrary trends of large wavelengths. In carbonates, it is useful to differentiate between more compacted limestones, with presence of clays, from those with good porosities.
- **Attenuation T***. It is an algorithm based on the attenuation of the frequencies that are above and below from the selected analysis points. In carbonates, it can indicate open fractures and detect the presence of gas.
- **Chaos**. It defines and maps how chaotic the local seismic signal is from the statistical analysis of estimates of dips and azimuth. It is used to identify reef stacks and for sketching of karsification and fractures.

As fluid indicators, it is recommended to use the attributes:

- **Structural Smoothing**. It simulates smoothing of data by local averaging with a weighted Gaussian filter. It can illuminate flat spots or contacts between fluids.
- **Instant Quality**. A quality factor is assigned from the instant envelope. It is used in the detection of fractures and can indicate the fluid content and the absorption character of the reservoir.
- **Genetic Inversion**. It is an inversion process similar to that of Neural Networks. It is a good indicator of hydrocarbons when combined with other inversion attributes such as the P wave versus S wave speed (PWV/SWV) or Poisson rate.

- **Dominant Frequency**. It is the square root of the sum of the squares of the instantaneous frequencies and the bandwidths. It reveals the variable spectral properties over time of seismic. Changes in the dominant frequency of some events may be indicative of the presence of hydrocarbons (Schlumberger 2012).

In the present research, the Relative Acoustic Impedance and Genetic Inversion attributes are used in order to create volumes of the physical properties, which are in the wells.

The Genetic Inversion attribute of the Petrel program uses two modern learning techniques: the Genetic Algorithm (GA) and the Neural Network (NN). Combined, these techniques provide the ability to generate, quickly, estimated petrophysical properties of seismic volumes or attributes with minimal data preparation, being essential to work both, the seismic data and the well, in the same domain. It is not necessary to define a low-frequency model or estimate a wavy. However, precise tie of wells and time–depth calibration are required (Schlumberger 2012). This allows the response obtained (inverted volume of a well-scale property) to best fit the seismic.

Genetic Inversion, although actually flexible, requires a correlation between well logging property and seismic data. For example, there is obviously a good correlation between the acoustic impedance measured in the well and the Relative Acoustic Impedance (RAI) calculated from the seismic amplitudes.

GA/NN algorithms implicitly handle many of the linear processing components, such as zero-phase data. The relative amplitudes must be preserved in the processing stream and the Amplitude Gain Control (AGC) should be avoided. Filters must suppress random and coherent noise. The relatively low seismic resolution and the high resolution of the well logs are solved by passing a low-pass filter to the well logs. The upper and lower limits of the learning area should be such that this interval is kept small, taking into account the fact that, for example, differential compaction can result in the same lithology having different seismic properties.

Genetic Inversion makes use of the following concepts:

- Selection: Similar to Charles Darwin's natural selection hypothesis, which favors only individuals best adapted to survive, in this case, those with the lowest errors are those selected.
- Crossover: In this step, chromosomes (combining weights) are exchanging genes (particular weight, within a combination) between them (the number of genes exchanged can be singular or multiple). This phenomenon occurs with a probability given after and during each iteration.
- Mutation: Like the theory of natural evolution, genes will be randomly replaced within the chromosomes. This ensures that the process does not converge to a local minimum. The probability of occurrence is a function of the iteration step; for example, the mutation is more likely to happen as soon as the evolution of the error function is reaching a plateau. However, in most cases, it is much lower than the crossover probability.

Within the complementary unconventional exploration methods, DEM is used for the detection of light hydrocarbons microseepage, fundamentally vertical. The applications for the offshore hydrocarbons exploration are based on the formation of columns of mineralized rocks above oil and gas deposits (Price 1985; Pardo and Rodríguez 2016). The hydrocarbons in migration are oxidized causing changes in pH/Eh, resulting in changes in mineral stability. These lead to the presence of carbonatization and silicification, which result in denser and erosion resistant surface materials (formation of positive residual geomorphic anomalies). These anomalies can also be triggered by structural events.

4.3 Materials and Methods

The DEM (90 × 90 m), used in this study, was taken from Sánchez Cruz et al. (2015), with source at: http://www.cgiar-csi.org/data/srtm-90m-digital-elevation+ETOPO2 database. The processing of these data consisted of the regional–residual separation of the submarine relief from the calculation of the Upward Analytical Continuation (UAC) to 500 m, subtracting this from the observed relief (DEM), thus allowing to emphasize residual (local) anomalies of lesser vertical amplitude.

The data of the drilled wells were taken from the EIUS Database (Pico Peña 2012).

The information of the seismic structures was taken from the exploration reports of Domínguez et al. (2016) and González et al. (2015).

The processing of the information was done using the Petrel Software version 2015 from Schlumberger.

Part of the automated processing and interpretation of the georeferenced geophysical–geological information, related to the morphotectonic regionalization, was carried out with the Oasis Montaj (R) System, version 7.01.

The morphotectonic regionalization was carried out by mapping the residual DEM to 500 m by plotting morphotectonic alignments, following the linearity, gradients, inflections, and interruptions of the contours, separating different tectono-structural elements, characterized by different levels of amplitude of positive residual anomalies.

In order to create volumes of physical properties that respond to fluids in the wells, the Genetic Inversion attribute (GI) and the training of estimation models using Neural Networks were used. The GI attribute is obtained on the Pre-Stack Depth Migration (PreSDM) and on the velocity information of the compressional (PWV—P wave velocity) and shear (SWV—S wave velocity) waves from the wells.

The first volume that was generated was the P wave Acoustic Impedance (PWAI). For this, it was necessary to edit the PWAI register of the well to reduce the resolution and to have a better adjustment with the seismic volume. The volume of the Relative Acoustic Impedance attribute was used, which gave better results since it is more related to the register. The Genetic Inversion algorithm was trained between two reservoir surfaces in order to better respond to areas with carbonate reservoir characteristics with fluids. Once the volume of PWAI was obtained, the volume of S wave Acoustic Impedance (SWAI) was calculated, the wells being adjusted to the RAI volume.

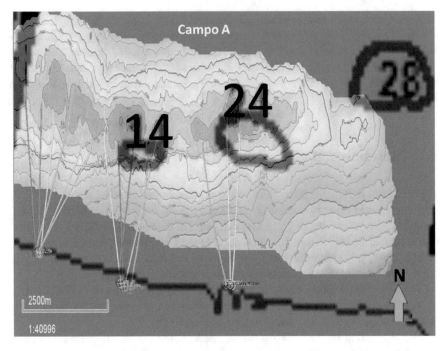

Fig. 4.1 Geomorphic anomalies (GAs) with amplitudes of 14 and 24 m associated with field A. Note drilled wells, oil producers, coinciding with two of the GAs

The next step was to calculate a volume of the ratio of compressional and shear velocities (PWV/SWV), for which an estimation model was used that uses Neural Networks to find a relation between the attributes calculated on the seismic volume and the register of well. The PWAI volume is divided between the SWAI volume, which results in a volume of PWV/SWV (since the densities in the numerator and the denominator are canceled), establishing the speed ratio, indicating the presence of reservoirs saturated with fluid.

4.4   Results

4.4.1   Comparative Analysis of the Geomorphic Anomalies Found in the Area

In the producer fields examined, the existence of a possible relationship between the amplitude of the geomorphic anomalies (GAs) and the depth of the producing structures was observed. Thus, in fields A and C (Figs. 4.1 and 4.2), structures with depths between −1500 and −1700 m (TVDSS—True Vertical Depth Sub-Sea)

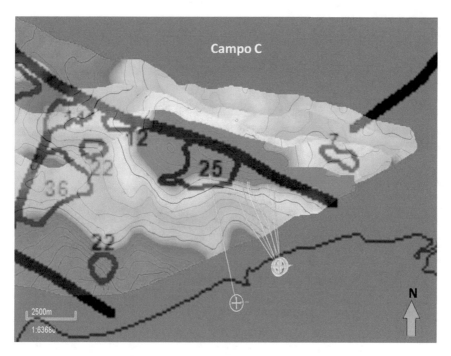

Fig. 4.2 Field C. It presents a GAs train with amplitudes between 12 and 25 m. Drilled wells coinciding with the 25 m GA are excellent oil producers

have anomalies that, although they differ about their organization and amount, have amplitudes of the same order (10–25 m). In field A (Fig. 4.1), a train of anomalies with values between 14 and 28 m associated with the interpreted structure is observed, although only those with amplitudes of 14 and 24 m have been drilled, as well as a part of the structure without anomalies, also with production. These anomalies are similar to those of field C (Fig. 4.2), which presents a GAs train with amplitudes between 12 and 25 m, only the latter having been drilled.

Another field examined was field B (Fig. 4.3). This is at depths greater than −1700 m TVDSS, and although it presents good oil production, the geomorphic anomalies (positive in character) are of amplitude negative and more dispersed, oscillating in the order of (−10 and −12 m). Also, a part of the structure without GAs has production.

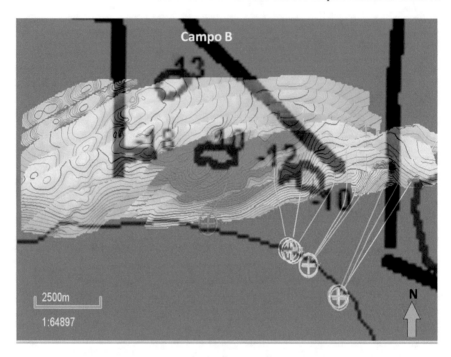

Fig. 4.3 GAs with negative amplitudes associated with the oil producing B field. To the east and north of this structure are anomalies of greater amplitude not yet drilled

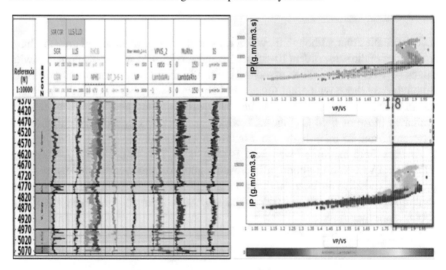

Fig. 4.4 On the left, the composite chart shows the seal zones in light blue. On the right, top and bottom, cross-graphs of the PWV/SWV–VP/VS and PWAI–IP ratio. They stand out with a red rectangle the reservoir areas

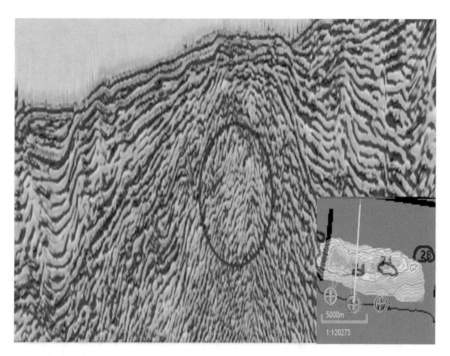

Fig. 4.5 Response of the GI attribute, marking an interval with velocity ratio values greater than 1.8, at a depth of 1100 m TVDSS within the regional seal, and associated with a GA of 14 m amplitude

4.4.2 Genetic Inversion and Its Relationship with Geomorphic Anomalies

The velocities of the seismic waves vary depending on the characteristics of the rocks that they cross, knowing that the compressional waves cross all the states of aggregation: solid, liquid, and gaseous, whereas the shear waves only propagate by the solid means. Therefore, when finding a rock with fluids the speed of the P waves is diminished whereas that of the S waves are not affected, since they cross through the skeleton of the rock. The PWV/SWV ratio is higher in the reservoirs, and the lower values are obtained in the seal, as a result of the well-scale analysis (Fig. 4.4) according to Reyes-Paredes and Fernández-Mederos (2017).

The evaluation of the Genetic Inversion seismic attribute (GI) allowed us to define the existence of saturated carbonate bodies with velocity (PWV/SWV) values higher than 1.8, indicating the existence of a saturated reservoir. In Fig. 4.5, in the field A, the existence of a possible reservoir within the regional seal, associated with a GA of 14 m of amplitude, is observed.

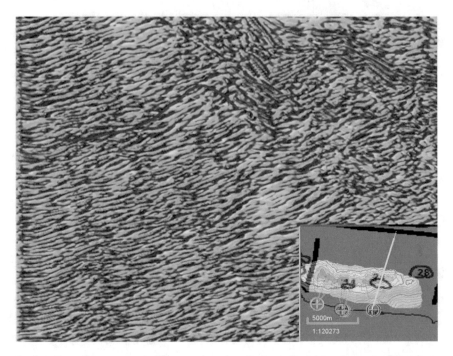

Fig. 4.6 Fractured carbonate body reservoir type, drilled to a depth of −1710 m TVDSS. In red, the top of the producing scale

Figure 4.6 shows the drilled structure at −1710 m TVDSS, where the response of the GI attribute verifies the existence of a reservoir associated with a geomorphic anomaly of 24 m of amplitude.

On the other hand, between fields B and C, one can observe the existence of geomorphic anomalies in non-drilled regions, where the Genetic Inversion attribute shows the existence of possible reservoirs (Fig. 4.7).

In another order, in Fig. 4.8, the response of the GI attribute to the fault zone and the possible extension of the field to the east of it is observed, as well as the existence of a structure to the north (orange circle).

4.5   Conclusions

There is an acceptable correspondence between the response of the GI attribute (which shows a permissible delimiter of the reservoir rocks) and the geomorphic anomalies. In general, the location, shape, and amplitude of the GAs correspond to the particular geological conditions of each area of study. As a rule, it shows a decrease in amplitude of the same for greater depths of the producing structures, and

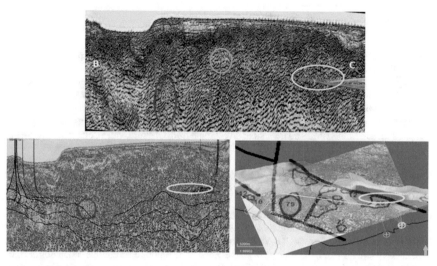

Fig. 4.7 Response of the GI attribute. Yellow circle highlighting the drilled zones with a GA of amplitude 25 m; in orange, zone not drilled with GAs of amplitudes of 22 and 36 m, indicating possible reservoir to a depth of −1212 m TVDSS; and in red, a fault zone to which an intense GA of 79 m is associated

Fig. 4.8 Fault zone associated with a GA of 79 m of amplitude is observed, and to the east of this, the possible extension of the reservoir body. To the north, a GA with amplitude of 17 m is observed, that is associated with a chain of anomalies (as it is observed in Fig. 4.6, box, bottom right)

anomalous positive values associated with fault zones. With the use of both methods, the possibility of defining new prospective areas for the search of hydrocarbons is proposed, so it is recommended to extend this type of study to other producing and potentially producing areas to confirm its validity.

References

Chopra S, Marfurt K (2006) Seismic Attributes—a promising aid for geological predicition. CSEG RECORDER Special Edition

Domínguez BR, González EM y otros (2016) Resultados de la exploración sísmica 3D y caracterización de prospectos exploratorios para petróleo y gas en los Bloques 7 y 7A, al Norte de La Habana y Mayabeque-Proyectos I + D 7080 y 9015. Ceinpet, La Habana, Cuba

Ekwe AC, Onuoha KM, Osayande N (2012) "Fluid and lithology discrimination using rock physics modelling and LambdaMuRho Inversion: an example from Onshore Nigel Delta, Nigeria" Search and Discovery Article #40865, Posted 30 Jan 2012

ETOPO 2. Fuente: http://worldgrids.org/doku.php/wiki:dem_sources

González EM, Reyes-Paredes O, et al (2015) Informe Parcial sobre la actualización del modelo geológico Varadero Oeste, Bahías de Matanzas y Cárdenas - Proyectos I + D 7080 y 9000. Ceinpet, La Habana, Cuba

Pardo Echarte ME, Rodríguez Morán O (2016) Unconventional methods for oil & gas exploration in Cuba. Springer Briefs in Earth Syst Sci. https://doi.org/10.1007/978-3-319-28017-2

Pedro-Pompa D, Reyes-Paredes O, Veiga-Bravo C, Miró-Pagés G (en prensa) Uso de atributos sísmicos y registros geofísicos de pozo para el análisis de fluidos en cortes carbonatados del Cinturón Plegado de la Provincia Norte Cubana. Revista Geociencias. Órgano de difusión científica de la Universidad Olmeca

Pico Peña R (2012) Sistema Único para la Información de la Exploración (SUIE)-Proyecto I + D 7038. Ceinpet, La Habana, Cuba

Price LC (1985) A critical overview of and proposed working model for hydrocarbon microseepage. US Department of the Interior Geological Survey. Open-File Report, no 85–271

Reyes-Paredes O, Fernández-Mederos JC (2017) Propiedades Acústicas en Reservorios Carbonatados de La Franja Norte de Crudos Pesados de Cuba. Memorias VII Convención de Ciencias de la Tierra, VI Congreso Cubano de Petróleo y Gas

Sánchez Cruz R, Mondelo F, et al (2015) Mapas Morfométricos de la República de Cuba para las Escalas 1: 1,000,000–50,000 como apoyo a la Interpretación Geofísica. Memorias VI Convención Cubana de Ciencias de la Tierra, VIII Congreso Cubano de Geofísica. Fuente: http://www.cgiar-csi.org/data/srtm-90m-digital-elevation+ETOPO2

Schlumberger (2012) Interpreter's guide to seismic attributes, 4th edn

Vargas-Meleza L, Megchun J, Vázquez G (2004) Petrophysical Properties Estimation by Integrating AVO, Seismic Inversion and Multiattribute Analysis in a 3-D Volume of Playuela, Veracruz. In: AAPG International Conference: 24–27 Oct, Cancún, México

Wang J, Dopkin D (2012) Visualization, analysis, and interpretation of seismic attributes for characterizing a carbonate reservoir. Fuente: https://www.spgindia.org/2008/375.pdf

Chapter 5
Use of the Digital Elevation Model for Offshore Morphotectonic Regionalization, in the Central East-North Cuba Sector

Abstract The tectonic regionalization of the marine territories for exploring oil and gas is a very current research problem. From the Digital Elevation Model (DEM) 90 × 90 m, in its residual variant at 500 m, the morphotectonic regionalization, offshore, in the Central East-North Cuba sector is performed. Four regions were identified (A, B, C, and D), where the tectonic environments of Platform (Remedios–Bahamas), Foreland Basin, and Thrust Belt are generally recognized. This regionalization was compared to the sismogeological scheme for the easternmost region of the area, serving as a general validation, even though with a greater degree of structural detail. It provides a terrestrial gravimetric regionalization scheme, which allows proposing a scheme of tectonic zonation of the territory, from its integration with the morphometric data. The results allow establishing a comparison between the anomalies derived from the seismic interpretation and the positive residual geomorphic anomalies of different amplitude, revealed by the residual DEM at 500 m. For the southeast of region C, the presence of small geomorphic anomalies with amplitudes between -40 and -10 m is established, and a smaller group with positive values in the first tens of meters. Correspondence with seismic anomalies is around 50%. In region D, the amplitudes 10–25 m predominate, with approximately 63% correspondence with the interpreted seismic anomalies. An evaluation of the degree of similarity between the section of the morphotectonic scheme belonging to the easternmost block and the section of the morphotectonic scheme corresponding to the Seboruco–Varadero, most productive sector of the North Cuban Oil Strip, is carried out. This allowed arguing, by simple analogy in the geological and morphometric data, the gasopetroliferous perspective of the most eastern sector of Cuba. It reaffirms the potential of the DEM (90 × 90 m) as a preliminary exploratory method that provides a rapid and effective form of research for offshore oil and gas.

Keywords Digital Elevation Model · Morphotectonic regionalization
Geomorphic anomalies · Seismic reflection · Seismic anomalies
North Cuban Oil Strip

© The Author(s) 2018   41
M. E. Pardo Echarte et al., *Offshore Exploration of Oil and Gas in Cuba*
using Digital Elevation Models (DEMs), SpringerBriefs in Earth System Sciences,
https://doi.org/10.1007/978-3-319-77155-7_5

5.1   Introduction

In recent years, there has been an increase in exploratory oil activity in the territory of the Bahamas with the detection of several prospects of interest for the location of deep wells. In this, the criteria proposed by López Rivera et al. (2012) on the possibility of deep and super-deep (saline, pre-salt) plays in the northeastern part of Cuba (land and sea) have been decisive.

In a first paper (Pardo Echarte et al. 2015), the use of the Digital Elevation Model (DEM) for the detection of subtle positive residual geomorphic anomalies possibly related to the hydrocarbons microseepage on offshore oil and gas targets is exemplified. It represents, for the first time, the use of this type of anomalies for exploratory purposes in marine territories. Hence, the implementation of complementary exploratory tools that allow a better evaluation of the prospective areas and, in turn, an increase of the geological effectiveness of the exploration justifies the use of the DEM (90 × 90 m), in the marine territory under study. This is located in the Exclusive Economic Zone (EEZ) Cuba–Bahamas–Haiti in the Central East-North Cuba sector (Fig. 5.1).

The aim of this chapter is to establish, based on the results of the DEM (90 × 90 m) and the corresponding morphotectonic regionalization, a validation and precision of the existing sismogeological scheme in the northeastern marine region of Cuba. This scheme has been obtained from seismic reflection data in the area, where the availability of seismic lines is poor and there are no offshore wells.

For reasons of confidentiality, maps are presented without coordinates.

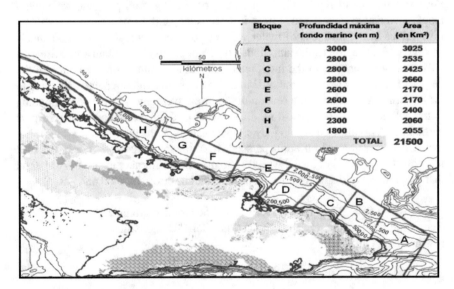

Bloque	Profundidad máxima fondo marino (en m)	Área (en Km²)
A	3000	3025
B	2800	2535
C	2800	2425
D	2800	2660
E	2600	2170
F	2600	2170
G	2500	2400
H	2300	2060
I	1800	2055
TOTAL		21500

Fig. 5.1 Blockade of the Exclusive Economic Zone (EEZ) Cuba–Bahamas–Haiti in the Central East-North Cuba sector. *Source* López Rivera et al. (2012)

The information provided by seismic investigations, north of the former Oriente province (Tecnoexport 1984–1987), indicates a geological constitution similar to that of the EEZ in the Gulf of Mexico; that is, the presence of thrust zones, to the south, foreland basin, to the center, and the Bahamas platform, to the north (López Rivera et al. 2012). The following is a brief synthesis of the constitution of these geological-tectonic units; from south to north:

- *Thrusted and Folded Belt of the North of Cuba (Rocks of the North American Continental Margin)*—Belt of rocks mainly sedimentary of Jurassic to Eocene age, with sublatitudinal course (NO-SE), that extends throughout the Cuban archipelago, in land and sea (western to eastern Cuba). These rocks have been distinctly deformed in a severe compressive environment.
- *Synorogenic Foreland Basin*—It extends along the northern coast of Cuba, between Mariel (western Cuba) and Gibara (eastern Cuba). It consists in a series of rock outcrops from the Paleocene to the Early Upper Eocene, which originally belonged to an extensive basin, at the head of the thrusts, located on the ProtoCaribe crust. Massifs of the Remedios Carbonate Bank (Bahamas Platform) sometimes interrupt them.
- *Bahamas Platform*—The rocks of its basement do not come out and are Paleozoic or older, according to deep drilling data. Igneous, metamorphic, and sedimentary rocks whose radiometric ages vary between 530 Ma (Cambrian) and 143 Ma (Lower Cretaceous Berriasian) represent them. In some areas of the territory, saline diapirs develop (García Sánchez 2017). Mainly massive limestones and dolomites integrate the sediments of platform.

On land, we can observe three other geological-tectonic units that have not been studied in the present work; from north to south:

- *Belt of Mafic-Ultramafic Rocks*—Integrated by rocks of the Ophiolitic Association and their serpentinitic "melánges," that occupy diverse tectonic positions, even on the Thrust Belt, or they are redeposited between sedimentary rocks of different geologic-tectonic units.
- *Cretaceous Volcanic Arc*—Composed of effusive, plutonic, and subvolcanic, volcanogenic-sedimentary volcanic rocks, as well as metasomatic and metamorphic rocks, ranging from the Hauterivian to the Middle Campanian (Iturralde-Vinent 1997), as well as post-volcanic synorogenic rocks with ages from the basal Paleocene (Danian) to the Middle Eocene (Lutetian).
- *Post-volcanic Synorogenic Basins*—Developed on the deformed remains of the Cretaceous and Paleogene volcanic arches (the latter, in the eastern provinces) and on the belt of mafic-ultramafic rocks, so they are basins entirely developed on the geological units of oceanic nature.

5.2   Theoretical Frame

The large-scale use of marine seismic data dates back to the 1960s, and its main impe-
tus was mainly due to its use in oil exploration. In the acquisition, along previously
established profiles, a group of seismic traces from all the shots is obtained, which
are processed and rearranged in sets of "common reflector points." These contain
the information of all the reflections found, from where it proceeds to group and sum
to obtain the reflection seismic section. The processing has vital aspects such as:
reduction of noise level, removal of static time variation, and analysis of the velocity
model for the sismogeological knowledge of the explored area. The interpretation
allows, with different levels of resolution, the cartography of different structures of
interest for gas–oil exploration. In the case of thrust belts, the seismic information
is ambiguous and confusing, where only border fragments can be visualized due
to the presence of very steep dips, complex, and variable velocity fields, as well as
frequent irregularities of the upper part of the section. This is not the case with basins
and anticlines, where, in general, there is a good seismic expression. The faults are
identified from divergences in the boundary dip, not associated with stratigraphy,
with notable changes in velocities to both parts of the seismic lines.

In marine territories, the coincidence of anomalies, geomorphic, and geophysical
allows to argue the combined use of unconventional methods of exploration, complementary to the seismic investigations. Within the complementary unconventional
methods of exploration, the DEM allows the detection of geomorphic witnesses of
light hydrocarbons microseepage, fundamentally vertical. Its applications for the
offshore hydrocarbons exploration are based on the formation of columns of mineralized rocks above oil and gas deposits (Price 1985; Pardo Echarte and Rodríguez
Morán 2016), linked to structural traps reflected by seismic. The hydrocarbons in
migration are oxidized causing changes in pH/Eh, resulting in changes in mineral
stability. These lead to the presence of secondary calcium carbonate mineralization
and silicification, which results in denser and erosion-resistant surface materials (formation of positive residual geomorphic anomalies). It should be pointed out that the
positive residual geomorphic anomalies could also have a lithological or structural
nature not linked to hydrocarbon microseepage processes.

5.3   Materials and Methods

The DEM (90 × 90 m) used in this work was acquired from Sánchez Cruz et al.
(2015), with source at: http://www.cgiar-csi.org/data/srtm-90m-digital-elevation
+ ETOPO 2. The processing of these data consisted in the regional-residual separation
of the relief from the calculation of the Upward Analytical Continuation (UAC) at
500 m, subtracting this from the observed relief (DEM), which allowed emphasizing
residual anomalies of lower vertical amplitude. The morphotectonic regionalization
was carried out from the map of residual DEM to 500 m by plotting morphotectonic

Table 5.1 Characteristics of marine seismic surveys in north-central eastern Cuba

Characteristics	1984	1986
Minimal distance	180 m	185 m
Maximum distance	2555 m	2560 m
Receptors		
Number of channels	96	96
Distance between channels	25 m	25 m
Number of receivers per group	24	24
Group basis	23	23
Sources		
Distance between shot points	50 m	50 m
Source type	PI-200	IGP-1
Volume of the source	$4 * 2.5 \, dm^3$	$4 * 3 \, dm^3$
Depth of the cable	7 m	8 m
Pressure	$120 \, kg/cm^3$	$120 \, kg/cm^3$
Estation	SERCEL SN388HR	SERCEL SN388HR

alignments following the linearity, gradients, deflections and interruption of the contours, and separating different tectono-structural elements characterized by different amplitude levels of positive residual anomalies.

The seismic lines used for integration with the residual DEM at 500 m were acquired in the seismic campaigns of Tecnoexport (1984–1987), whose data are appended in Table 5.1. The Geophysics Unit of Science and Technology of Base (USTB) of the institution reprocessed them in the year 2016, with the program GeoEast, version 2.6.3, of Chinese origin. The result was the migration in time after the sum in the mentioned lines (Lastra Cunill et al. 2016).

The gravimetric regionalization of the northern terrestrial territory was carried out from the grid of the gravimetric field (Bouguer, 2.3 t/m³), at a scale of 1:250,000, taken from Mondelo et al. (2011). The processing of the gravimetric data consisted in the obtaining of the total horizontal derivative for the tracing of tectonic alignments, following the axes of the maxima and, considering the abrupt deflections and interruption of the characteristics and levels of the field that separate different elements or tectono-structural units.

Much of the automated processing and interpretation of the georeferenced geophysical–geological information, related to morphotectonic and gravimetric regionalization, was carried out with the Oasis Montaj System (R), version 7.01. The remaining processing was done with the Schlumberger Petrel System, version 2015.

5.4   Results

5.4.1   Morphotectonic Regionalization Versus Seismic-Geological Zonation of the Region Central East-North Cuba

The residual DEM at 500 m and its corresponding morphotectonic regionalization are presented in Figs. 5.2 and 5.3, respectively. According to the level and behavior of the residual relief field, four regions can be identified: A, B, C, and D. Regions A and C, with a level between 70 and −15 m and a quiet behavior, are associated to a tectonic environment of Platform (Remedios–Bahamas). Region B, characterized by a level of −15 to −70 m and a variable behavior, is identified with a tectonically depressed zone ("Foreland Basin") bordering to the north with the Bahamas Platform. The region D, characterized by a level of 15 to −15 m and a variable behavior, is associated to the Thrusted and Folded Belt of the north of Cuba, limiting, to the north, with the Foreland Basin.

Although the exploratory seismic level of the territory is low, the real possibility of discovering oil and gas deposits in the northern seas of the most eastern provinces of Cuba is based on the results of different investigations, both in the reference sector as well in their environment (López Rivera et al. 2012). In this way, the probable presence of active oil systems and a tectonic-structural zonation in the northeastern seas, from Nuevitas to Maisí, has been established. It consists of the Thrust Belt, the Foreland Basin, and the Bahamas Platform. It should be noted that basin-platform relationships are, from the point of view of oil potential, more complex but perhaps more favorable. Thus, according to López Rivera et al. (2012), it is of interest the west area of Gibara–Nuevitas, with the location of the Remedios Platform, to the

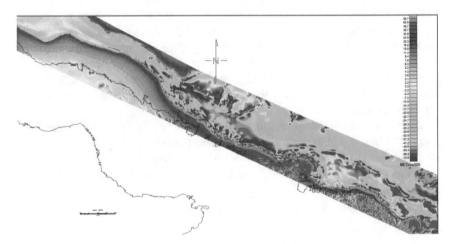

Fig. 5.2 Residual DEM at 500 m from the north-central east marine territory of Cuba

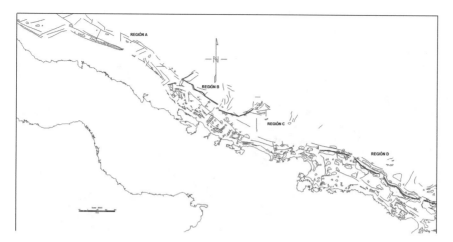

Fig. 5.3 Morphotectonic regionalization, with positive residual geomorphic anomalies, of the north-central east marine territory of Cuba. In red, first-order morphotectonic alignments

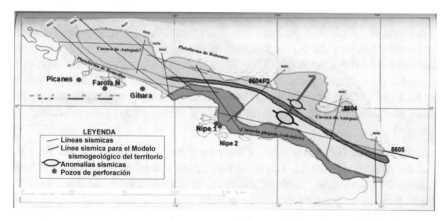

Fig. 5.4 Sismogeological Zoning Scheme of the marine territory of Central East-North Cuba, according to López Rivera et al. (2012), scale 1:500,000 in geographic coordinates

south (according to the seismic interpretation), a depressed area of deep water in the center (Foreland Basin?) and the Bahamas Platform, to the north. On the other hand, the displacement toward the seas of the north of Eastern Cuba (east of Gibara), due to transcurrent faults, of the Regional Belt of Northern Cuban Gravimetric Minimums, has been established. This belt of minimums is associated to the Thrust Belt (due to the presence of great sedimentary thicknesses). The results of the sismogeological interpretation are shown in Fig. 5.4.

The gravimetric regionalization of the northern terrestrial territory of Central East Cuba was proceeded. The results of this regionalization are given in Fig. 5.5. A different interpretation is observed: The coastal area between the transverse tectonic line

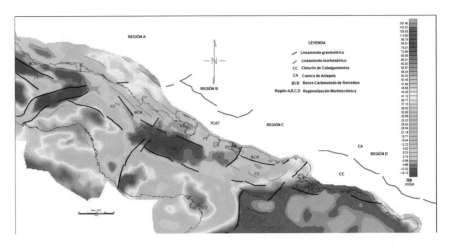

Fig. 5.5 Scheme of gravimetric regionalization of the terrestrial territory of Central East-North Cuba

to the west of Gibara and the transverse fault that follows it (Camagüey) is characterized by low-amplitude gravimetric minimums, presumably associated with the Foreland Basin (FB-CA), north of the Thrust Belt. This interpretation is supported by another one, which involves terrestrial seismic and drilling data (Valdivia-Tabares 2015). The Foreland Basin seems to extend uninterruptedly, toward the sea, to the limit with the Bahamas Platform, to the north (region B of the morphotectonic regionalization).

- In the areas of Gibara the west of the transverse fault Camagüey and the west of the cross-fault zone La Trocha-Tuinicu, the rocks of the Remedios Platform (Remedios Carbonated Bank-BCR) appear in tectonic subvertical contact (>70°) with those corresponding to the Thrust Belt.
- In such localities, the rocks of the FB-CA could have been squeezed and, perhaps, partially eroded, by the violent compression. To the north of the last mentioned zones is the Bahamas Platform (regions, A and C, of the morphotectonic regionalization). The minimums belt of the gravimetric field, limited by tectonic alignments, maps the Thrust Belt (CC), of sublatitudinal course. It is displaced toward the northern seas, east of Gibara, by the easternmost transverse tectonic line (Cauto-Nipe fault zone).

As a synthesis of the above information, it was integrated into a Tectonic Zoning Scheme, shown in Fig. 5.6.

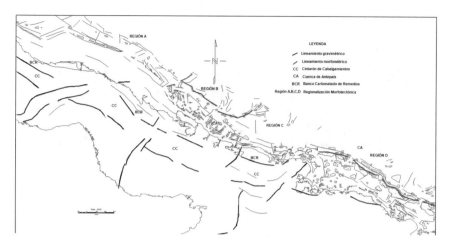

Fig. 5.6 Tectonic Zonation Scheme of the Central East-North Cuba, where terrestrial gravimetric and offshore morphotectonic regionalization are integrated

5.4.2 Comparative Analysis Between the Morphotectonic Scheme with Positive Residual Geomorphic Anomalies Versus the Mapped Seismic Structures

Given the novelty in the use of DEM as a method to detect offshore anomalies possibly related to oil and gas targets, a comparison is made between positive residual geomorphic anomalies of different amplitude (in blue) and the anomalies resulting from the seismic interpretations (in black) (López Rivera et al. 2012; Gómez 2017) (Fig. 5.7).

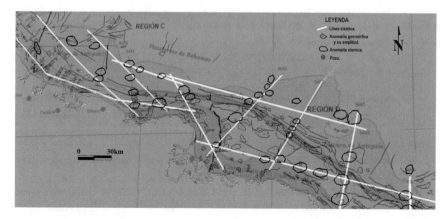

Fig. 5.7 Integration of positive residual geomorphic anomalies of the residual DEM at 500 m (in blue) with seismic anomalies (in black)

(a)   **(b)**

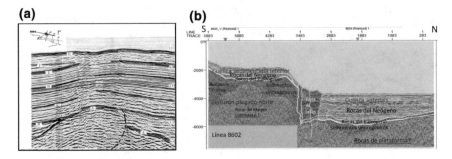

Fig. 5.8 a Sector of seismic line 8604: anomaly with apparent characteristics of a gas chimney presence. **b** Proposal of the Sismogeological Model of the area, according to the seismic interpretation of Line 8602

Because of the availability in the interpretation, only of the seismic lines located in the easternmost part of the territory, the comparison focuses, exclusively, on the corresponding area between regions C and D of the morphotectonic regionalization scheme.

From the morphometric point of view, the presence of several small anomalies with an amplitude between −40 and −10 m and, a smaller group with positive values in the first tens of meters, is established for the southeast portion of region C. The correspondence with the seismic anomalies is around 50%. In region D, entering the Thrusted and Folded Belt, there is an increase in the presence of geomorphic anomalies, with an increase in the size and amplitude of the same, predominating the level 10–25 m. There is approximately 63% of correspondence with the seismic anomalies interpreted.

Some results that exemplify the seismic interpretation of the territory are shown in Fig. 5.8: (a) On line 8604, near the intersection with line 8601, an area is observed whose characteristics suggest the presence of a gas chimney. This would confirm the extension of the oil systems present on land to the marine territory (López Rivera et al. 2012). (b) It presents the sismogeological model of the territory, obtained from recent interpretations (Gómez 2017).

In Fig. 5.8b, the characteristics of the seismic wave field for each identified geological-tectonic unit have been defined (Gómez 2017):

- **Thrusted and Folded Belt**: seismic horizons of mean frequencies and amplitudes, which are interrupted at short distances. Their arrangement is, generally, with different angles, resembling the geometry of thrusted or scaled folds. It is a "noisy" environment and has 3D influence.
- **Foreland Basin**: long and well-expressed dynamically seismic horizons. Sismostratigraphic discordances are also sometimes present.
- **Platform (Remedios–Bahamas)**: seismic horizons of variable frequencies and amplitudes with losses of regular energies, due to the presence of dense carbonates (Miró-Pagés 2008).
- **Inland Sedimentary Basin**: expressive horizons, generally, of good resolution.
- **Fault Zone**: seismic horizons that are interrupted in the same fault plane.

5.4.3 Comparative Analysis of the Easternmost Section of the Morphotectonic Scheme Presented and the One Corresponding to the Seboruco–Varadero Productive Sector in the Cuban North Oil Strip

A comparative analysis is established on the distribution and type of positive residual geomorphic anomalies resulting from the residual DEM at 500 m, between the western part of Region D of the presented morphotectonic scheme and that corresponding to Region B of the Habana-Matanzas territory, coinciding with the Seboruco–Varadero productive sector of the NCOS (Fig. 5.9).

In this regard, a zonation of west Region D of Central East-North Cuba has been established, according to the size, amplitude, and type of distribution of positive geomorphic anomalies, recognizing four main areas of sublatitudinal direction. Of interest, for the purposes of the comparison with region B of Habana–Matanzas, zones one and two results. In both, predominate anomalies with amplitudes between 10 and 25 m, as in the case of Region B. In addition, for zone 1, near the coast, the smaller and randomly distributed anomalies predominate for the two regions. For zone 2 (10–20 km of the coast), the anomalies of both regions are of greater proportions and have a greater tendency to form chains.

Taking into account that both regions are located on the Thrusted and Folded Belt and, given the existence of hydrocarbon shows in the eastern coastal zone; recognizing the morphometric similarity noted above for both regions (D and B), this justifies proposing, a priori, the gasopetroliferous perspectives of Cuba's most eastern marine sector.

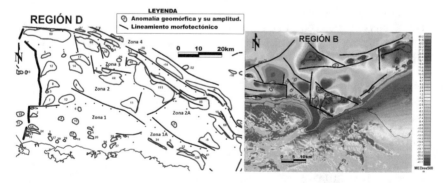

Fig. 5.9 Establishment of morphometric similarity between Region D (west) of the territory of Central East-North Cuba and Region B of the territory of Habana–Matanzas

5.5   Conclusions

- From the Digital Elevation Model (DEM) 90 × 90 m in its residual variant at 500 m, the offshore morphotectonic regionalization, is performed in the Central East-North Cuba sector. It was divided into four regions (A, B, C, and D), where the tectonic environments of Platform (Remedios–Bahamas), Foreland Basin, and Thrust Belt are recognized. This regionalization was compared with the existing sismogeological scheme for the easternmost region of the area (based on reflection seismic data), serving as a general validation, even though with a greater degree of structural detail. It was argued about the preparation of a terrestrial gravimetric regionalization scheme, which allowed proposing a scheme of tectonic zonation of the territory, from its integration with the morphometric data.
- The results allowed to establish a comparison between the anomalies derived from the seismic interpretation and the positive residual geomorphic anomalies of different amplitude, revealed by the residual DEM at 500 m. For the southeastern part of the region C, the presence of small geomorphic anomalies with amplitudes between −40 and −10 m, and a smaller group with positive values in the first tens of meters, was established. The correspondence with the positive seismic anomalies was around 50%. In the region D, in the Thrusted and Folded Belt, there is an increase in the presence of geomorphic anomalies, with an increase in the size and amplitude of the same, predominating the level 10–25 m. There is approximately 63% of correspondence with the seismic anomalies interpreted.
- The degree of similarity between the section of the morphotectonic scheme belonging to the easternmost block (east of Holguín) and the section of the morphotectonic scheme corresponding to the Seboruco–Varadero sector of the NCOS, in Habana–Matanzas, allowed to argue, a priori, the gasopetroliferous perspective of the most eastern sector of Cuba.
- The results obtained confirm the potential of the DEM (90 × 90 m) as a preliminary exploratory method, which provides a fast and effective form of research for offshore oil and gas.

References

ETOPO 2. Fuente: http://worldgrids.org/doku.php/wiki:dem_sources

García Sánchez R (2017) Geología Regional. En: Caracterización geológica de la región Caribe – Mexicana, Inédito. CUJAE, 55 pp

Gómez I (2017) Modelo sismo-geológico en un sector marino de aguas someras de la región norte oriental cubana, Inédito. CUJAE, 37 pp

Iturralde-Vinent MA (1997) Introducción a la geología de Cuba. In: Furrazola-BermúdezG, Nuñez Cambra K (eds) Estudios sobre la geología de Cuba. Centro Nacional de Información Geológica, La Habana, pp 35–68

Lastra-Cunill M, Díaz-Fonseca R, Gómez R, González B (2016) Informe sobre el procesamiento de las líneas del sector centro-oriental adquiridas en los años 1984 y 1986. Inédito. Archivo CEINPET, La Habana, 78 pp

López Rivera JG, Morales-Carrillo I, Delgado O, Domínguez BR, Mejías L, Sosa C, Sánchez J, Linares E, Pról JL, Fernández PA (2012) Apertura a la exploración petrolera del sector Cuba – Bahamas – Haití de la Zona Económica Exclusiva de Cuba. Inédito. Archivo CEINPET, La Habana, 93 pp

Miró-Pagés G (2008) Curso de Posgrado de Prospección sísmica. En: La interpretación sísmica, Inédito. CUJAE, 45 pp

Mondelo F, Sánchez R et al (2011) Mapas geofísicos regionales de gravimetría, magnetometría, intensidad y espectrometría gamma de la República de Cuba, escalas 1:2 000 000 hasta 1:50 000. Inédito. IGP, La Habana, 278p

Pardo Echarte ME, Rodríguez Morán O (2016) Unconventional methods for oil & gas exploration in Cuba. Springer Briefs in Earth System Sciences, https://doi.org/10.1007/978-3-319-28017-2

Pardo Echarte ME, Súarez-Leyva V, Prieto-Castro F (2015) Empleo del Modelo de Elevación Digital (MED) para la exploración de hidrocarburos costa afuera. Inédito. Ceinpet. La Habana, 13 pp

Price LC (1985) A critical overview of and proposed working model for hydrocarbon microseepage. US Department of the Interior Geological Survey. Open-File Report 85-271

Sánchez-Cruz R, Mondelo F et al (2015) Mapas Morfométricos de la República de Cuba para las escalas 1:1 000 000–1:50 000 como apoyo a la Interpretación Geofísica. Memorias VI Convención Cubana de Ciencias de la Tierra, VIII Congreso Cubano de Geofísica. Fuente: http://www.cgiar-csi.org/data/srtm-90m-digital-elevation + ETOPO 2

Tecnoexport (1987) Complejo de trabajos geofísicos regionales marinos en el Shelf de Cuba centro-oriental. Informe de procesamiento de las campañas de 1984-87. Contrato 55-044/ 87000.Inédito. Ceinpet. La Habana, 115 pp

Valdivia-Tabares CM (2015) Informe de resultados de la evaluación del potencial de hidrocarburos del Bloque 17. Inédito. Archivo Ceinpet, La Habana, 84 pp

Printed in the United States
By Bookmasters